Table of Contents

1. Getting Started with Mathcad   1
   1.   The Mathcad Environment   2
   2.   The Handbooks   10
   3.   Formatting: Numbers, Graphs, Text   13

2. Basic Algebra   25
   1.   Algebraic Expressions and Maple   26
   2.   Factoring Using Symbolic Algebra   32
   3.   Solving Equations   38

3. Functions and Graphs   43
   1.   Functions, Graphs and Tables   44
   2.   The Root Function and Solve Blocks   48
   3.   Exponential Growth and Decay   53
   4.   Graphical Formats: Logarithmic   57
   5.   Polynomial Functions   61
   6.   Graphing Singularities and Discontinuities   65
   7.   Conic Sections   70
   8.   Parametric Curves and Polar Coordinates   77

4. Systems of Linear Equations   83
   1.   Graphical and Numerical Solution Techniques   84
   2.   The Use of Determinants   89
   3.   Matrix Methods for Systems of Equations   94

5. Statistics   100
   1.   Measures of Central Tendency and Variance   101
   2.   Linear Regression and Interpolation   109

6. Trigonometric Functions   114
   1.   Trigonometric Functions, Their Reciprocals and Inverses   115
   2.   Trigonometric Equations   124

7. Complex Numbers   126
   1.   The Use of Complex Numbers   127

8. Calculus of One Variable   132
   1.   Evaluating Limits Using Sandwiches   133
   2.   Derivatives of Polynomial Functions   137
   3.   The Root Function: Secant and Newton Methods   142
   4.   Critical Points in Polynomial Functions   148
   5.   Derivatives of Trigonometric Functions   153
   6.   Derivatives of Exponential and Logarithmic Functions   157
   7.   Definite Integrals and the Area Beneath the Curve   161
   8.   Numerical and Symbolic Integration Techniques   166
   9.   Applications of the Definite Integral   172

9. Series Approximations   176
   1.   Taylor Series   177
   2.   Fourier Series and Smoothing Functions   185
   3.   Fourier Sampling and the Analysis of Coefficients   194
   4.   Fast Fourier Transform and Data File Structures   199

10. Linear Differential Equations   205
   1.   Euler's Methods in Differential Equations   206
   2.   The Laplace Transform   214

Appendices   223
   1.   Coordinate Axis Rotation   224
   2.   Linear Regression Theory   226

Preface

This book is intended primarily for students of college technology, science and engineering programs who need to complement their basic mathematical understanding with a tool allowing ready visualization and flexibility. The content also applies to any general level functions and relations, pre-calculus and calculus courses. The book will appeal especially to those who enjoy learning mathematics through discovery and to those who, as yet, perceive mathematics as unnecessarily rigid and unforgiving.

Mathcad is a professional tool of visualization, simulation and problem solving. It allows complementary representations of processes through a multitude of built-in features and operators. Mathcad holds many advantages over other mathematics software tools. It is extremely accessible, does not suffer from the burden of unfamiliar syntax and has a familiar interface. Mathcad offers a ready, yet powerful, tool for the inquiring mind.

In each section of the text, a short introduction orients the reader to the general subject matter. This is followed by the Warmup, exercises and discussions of the applicable Mathcad techniques designed to be worked through and extended. At the end of each section, the Explorations provide the student an opportunity to apply the techniques to the solution of a problem or to the discovery of a mathematical pattern. Exercises within Explorations/*The Basics* include direct applications of Mathcad to a particular problem and may also include some conventional-style textbook exercises, while the problems within Explorations/*Beyond the Basics* offer an opportunity for open-ended problem solving and experimentation.

The use of any tool that allows mathematical exploration may at first be disorienting. After all, mathematics is one of the hard sciences with well-defined problems and equally well-defined solutions (the right answer!). Engineering models that use 'right answers' are responsible for all of our technologies. However, software tools also allow the engineer to play with the design and to experiment with alternate strategies and new possibilities.

Don't rush through these problems. Take your time. Documentation encourages your full attention to the problem. Students have told me that they only truly understood the significance of a problem and its solution after having written the documentation. Pay *full* (this is the catch!) attention to the complete process associated with one problem and you understand all like problems. You also will have developed a keener mathematical sense and will have a greater set of tools applicable to other problems. The foundation you dig with attention is deep.

The use of Mathcad produced unexpected results in my students. They openly discussed problems and solutions and moved naturally into group work within the laboratory setting. My role changed from that of teacher to that of facilitator. The greatest pleasure, however, was in having to remind students that the *mathematics* class was *over*.

Mathcad has provided me with a tool of play and discovery. I hope you find, with the help of this text, that it allows you an equal opportunity.

Supplement

A supplementary disk containing selected solutions is available to course instructors who adopt this manual.

Acknowledgments

I would like to thank Ron Doleman, Publisher, for his direction, support and humanity. Abdul Abdullahi, Editorial Assistant, provided indispensable coordination of the initial reviews. The constructive suggestions of the many reviewers have been gratefully incorporated within the text. Unbounded thanks are due, in memory, to my parents who imparted and instilled a love of numbers and learning. And many thanks to the students who completed exercises from the forerunners of this text and especially to those whose work exceeded my expectations.

Dedicated to Ellen, Oliver and Emma

E ... ns

M ... D

KEVIN P. DesRUES

Addison-Wesley Publishers Limited

Don Mills, Ontario • Reading, Massachusetts
Menlo Park, California • New York • Wokingham, England
Amsterdam • Bonn • Sydney • Singapore • Tokyo • Madrid
San Juan • Pari... ...City • Taipei

Publisher: Ron Doleman
Managing Editor: Linda Scott
Copy Editor: Rob Glen
Production Coordinator: Wendy Moran
Manufacturing Coordinator: Sharon Latta Paterson
Cover Design: Anthony Leung

Mathcad is a trademark of MathSoft, Inc.

Canadian Cataloguing in Publication Data

DesRues, Kevin
    Explorations in Mathcad

ISBN 0-201-42792-3

1. MathCAD (Computer file).   2. Mathematics—Data processing.
I. Title.

QA76.95.D47 1996   510′285′5369   C96-932238-0

ISBN 0-201-42792-3

Printed and bound in Canada.

A B C D E -MP- 01 00 99 98 97

Chapter 1: Getting Started with Mathcad

In this chapter, the Mathcad environment is introduced. The creation of text regions and areas of mathematical definition is examined. Equations are solved based on the definition of the included variables. Tables of values are generated for functions. As well, a reference feature of Mathcad, Electronic Handbooks, is examined.

Mathcad, in all its DOS and Windows versions, offers an incredibly rich environment for the solution of mathematical problems. Strategies may be approached numerically, analytically and/or graphically. Comparisons between the usefulness of these parallel methods can be readily created. As a numerical processor, Mathcad is able to return answers quickly and in a personally editable format.

More important than its use in the strict solution of problems is its application in the creation of an environment suitable to exploration and learning. Many educational analysts feel that the learning process involves a great deal of play, at any level and at any age. Too often, mathematics is perceived, taught and learned as answer and technique driven. The development of the solution method, with all its inherent successes and dead ends, is forgotten. Or, at best, invisible. An all-or-nothing approach is created which while useful in getting immediate answers belies the opportunities for creativity and discovery which have always been a part of mathematics.

Although there exist far more powerful mathematical software programs, Mathcad's advantage lies in the elegance of its interface and its ease of use. The metaphor used is that of a blank page waiting to be 'written' upon. You scratch out solutions, mixing text references and explanation with more mathematically active regions of tables, constants, equations, functions and graphs.

Mathcad does not solve the problems for you. Nor is it a tutorial tool although it can be used as one with the proper guidance. It is a professional quality tool used by thousands of scientists and engineers in the visualization and application of mathematics. It demands an understanding of mathematical syntax and grammar and rewards you by taking away the need for painful and tedious calculations. Time usually wasted repeating a technique endlessly is freed up to examine and explore.

Mathcad allows you to create spreadsheet type files which are forever and highly editable.

1.1 The Mathcad Environment

When Mathcad is loaded into your computer, the first screen contains many icons and a menu bar across the top of the page. Figure 1 is a screen capture of a Mathcad PLUS 6.0 empty worksheet. Notice that within the working area there are no directions, no hints but merely empty space. Mathcad uses the metaphor of the blank page to set the tone of its environment. On a paper page, you can cut and paste, move and erase anything you write. You may change the order of the content and yet still be able to see everything on the page. Mathcad offers this same flexibility.

Other mathematics software packages may fix the particular regions (text, calculation, tables, plots) to certain areas of the screen or even run particular data types in separate windows. However, with Mathcad everything is "on the page". The most searching you may have to do is accomplished by scrolling the page (which can become a problem on its own depending on your hardware and the size of your worksheet).

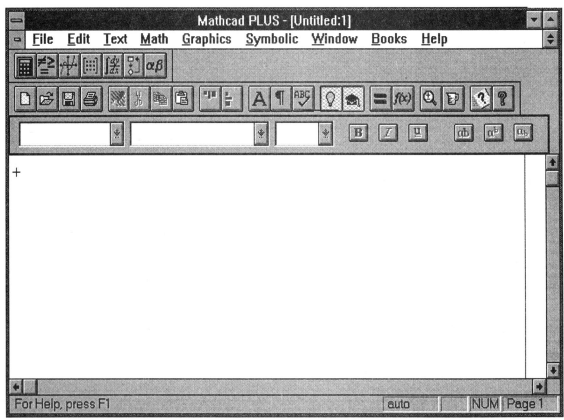

Figure 1

Although Chapter 1 is not strictly a set of mathematical explorations, the sections provide an introduction into some of the syntax and design of Mathcad. We will examine the creation and editing of an equation, the definition of variables, the opening and saving of files, the creation of functions and tables, and a subset of a symbolic processor (Maple) included with Mathcad.

> **Warmup**

Load the Mathcad software by double-clicking on its icon in the Program Manager screen (in Windows 3.1 or Windows 3.11) within the Mathsoft Group.

Sample Files

To examine and interact with ready-made examples of Mathcad files, open a file by:
1. clicking on the Open File icon from the Tool Bar (horizontal icon bar) or
2. selecting File/Open... from the pull-down menu bar , or
3. pressing F5, or
4. using the keyboard shortcut CTRL-O, or
5. select QuickSheets from the Help menu in Mathcad 6.0.

☑ As with all Windows programs there are many parallel ways of accomplishing the same task. In this text, the mouse method will usually be given followed by the pull-down menu and keyboard shortcut methods.

☑ Unless otherwise noted, 'clicking' refers to the use of the left-hand mouse button. The Mathcad 6.0 program refers to the row of graphical, calculus, Boolean and programming tools as the Palette, the row of icons as the Tool Bar, and the row of text formatting information as the Font Bar.

☑ Selections from the pull-down menus will be represented by their successive selections. For example, choosing to close a file will be represented as "select File/Close".

By default, Mathcad will go to the sample files included within the program. These files are part of the drive_letter:\Winmcad\ directory on your hard drive and are easily identified by the extension *.MCD. You may open as many of these at the same time as memory allows.

Once you have examined and edited the files, you may save them by:
1. clicking on the Save to File icon from the Tool Bar, or
2. selecting Save from the File pull-down menu, or
3. pressing F6, or
4. using the keyboard shortcut CTRL-S.

To begin, let's work with the creation of a new file. Select New from the File menu to create a blank worksheet.

A New File: Simple Calculation

Click somewhere in the empty region. A crosshair should appear. This indicates the start of the mathematically active region. Type "3.45 + 6.34 =" (note: without the quotation marks). Now move the cursor out of the number region by either clicking the mouse somewhere else on the page or by using the SHIFT-ENTER key combination.

The answer of 9.79 will appear automatically beside the ' = '. If the answer region stays blank, you may have loaded Mathcad with the Automatic Mode option turned off. If this is the case, a CALC-F9 reminder will appear in the lower right-hand corner of your screen. Pressing F9 then forces the program to perform the calculation at your convenience. If you want Mathcad to continuously update the answers as you create your solutions, then make sure the Math/Automatic Mode option is selected (i.e. has a tick mark beside it).

☑ When involved in lengthy editing sessions for large documents, turn the
Automatic Mode off. The distraction of having Mathcad track your every
editing decision is time-wasting and annoying. If you require a recalculation, a
press of F9 will then start the evaluation of the document up to the last line in
your current screen. If you want the calculation to extend to the whole of a
lengthy document, then select Math/Calculate Document.

☑ If, at any time, you forget a particular command or key, select Help or press F1
to access the general system help files.

Given all the time taken to load the program and type the numbers, this is an intentional misuse
of the program. Better to slice bread with a chainsaw. The tool used should match the scale of the
problem it was designed to solve.

Arithmetic and Editing with the Notched Rectangle

The arithmetic operations are accomplished by using the +, - , * and / keys for addition,
subtraction, multiplication and division respectively. We will apply the latter operations to the
two numbers we previously added. This will give us a chance to use Mathcad's editing features.
These features make heavy use of the spacebar and the up- and down-arrows and only *seem*
intuitive after you have worked with them for years!.

Move the cursor back into the region containing numbers. Click on the region near a number. A
vertical editing line should appear. A press of the backspace key will erase the character to the
left of the line. Now move the cursor out of the region and click. The numbers should be clear
once again and the new answer listed.

Note that the editing line cannot be applied to the result on the right-hand side of the equality.

To edit the operators, move the cursor near the + sign and click. A notched rectangular box
should appear which includes the two numbers and the operation but excludes the equals sign.
The same selection could have been accomplished by first clicking near a number to invoke the
vertical editing line then pressing the up-arrow key (or the spacebar) until the notched selection
box expands to contain the expression.

Press the backspace key.

The operator disappears and is replaced by an empty placeholder. Type in the operator you wish
included (- , *, or /). If in Automatic Mode, move the cursor outside your edited region to see
the result. Otherwise, press F9.

The notched selection rectangle is also useful for Copying, Cutting and Pasting expressions. The
repeated expressions of Figure 2 were created by first selecting the additive expression, then
selecting Edit/Copy (or CTRL-C) and pasting (select Edit/Paste or press CTRL-V) the expression
three times below the original. Each new expression was then edited for the inclusion of a new
operator.

The size of the notched rectangle is controlled by the repeated presses of the up- or down-arrows.
As the up-arrow (or spacebar) is pressed, the rectangle grows to include more and more of the
expression until the whole expression is contained. Repeated down-arrow presses decrease the
extent of the rectangle and end in the vertical editing line.

The direction of the notch indicates the direction of the editing. Inputs (operators and numbers) are placed to the right of the existing expression if the notch is in the top right-hand corner. To alter the notch direction, press the Insert key after creating a selection. The notch will migrate to the top left-hand corner and numbers and operations will be placed ahead of the selection rectangle.

Try the following:
- in a blank area of the worksheet, type 4.58
- press the ' (apostrophe) key. The number will be surrounded by brackets.
- press the up-arrow or spacebar twice to select the whole expression
- press the Insert key
- type "cos"
- press =

The equation cos(4.58) = -0.132 appears on the worksheet.

Text Regions

The text regions to the right of the mathematics area were created by clicking in an empty region, then by typing the double-quote ("). Quote regions may also be created by selecting Text/Create Text Region. A text box appears into which you can insert notes or meaningful explanations. To move out of the text region, click the mouse in an empty region or press Shift-Enter. Various icons beside the font selection windows at the top of the work page are used for formatting these text regions.

The Dashed-Line Rectangle

Another selection tool exists which allows to select and drag various types of regions (text, graph, table, definition) from place to place on the worksheet. Click outside the area you wish to select and hold the mouse-key down. This action fixes a corner of the selection rectangle. As you drag the mouse across and down (or up) a dashed-line rectangle will form which will include the areas you wish to move. Release the mouse-key. As you move the cursor back into the selected region, a large cross appears. Click and drag the regions to their new locations.

If this process jumbles up your content, select Edit/Regions/Separate (or CTRL-S) to separate overlapping regions and CTRL-R to refresh the screen.

$3.45 + 6.34 = 9.79$	*example of addition*
$3.45 - 6.34 = -2.89$	*example of subtraction*
$3.45 \cdot 6.34 = 21.873$	*example of multiplication*
$\dfrac{3.45}{6.34} = 0.544$	*example of division*

Figure 2

Obviously a hand-held calculator would be more use for certain operations.

Save your first Mathcad file by clicking on the Save File icon, by selecting File/Save, pressing F6 or CTRL-S. Choose a filename which indicates, to some degree, the purpose of the file. Hunting through 50 files named mcad01.mcd to mcad50.mcd is a self-defeating exercise.

Definitions and the Creation of Expressions

For this section, choose to create a new file (select File/New) or to simply extend your existing (and previously saved) file.

Let's examine the creation of an expression with all its variables and constants. For the evaluation of the equation $d = \frac{(a+b)^2}{c} + \sqrt{e} \cdot f$ for certain values of *a, b, c, e,* and *f* we need to first define these to have definite values. The equation is then defined and finally evaluated by substitution.

Although each variable is defined to be only one number, the expressions are editable and can be varied at will. Figure 3 shows the definition of each of the variables, the equation itself and the answer.

The definition symbol ': = ' is created by *typing* " : " (the colon sign) while the evaluation symbol is created by typing " = ".

The creation of the equation follows the following recipe:
- click on an open region
- type d : (the definition symbol will now appear followed by a small notched rectangle)
- type (a + b) ^ 2
- press the up-arrow (or the space-bar) once to select the squared parentheses
- press / and type c
- press the up-arrow once to select the whole fractional expression
- type + \ e (up-arrow to select all of the squareroot expression) * f

To evaluate the equation given the values of the variables, click on an open region to the right or below the equation's definition (Mathcad reads from top to bottom, left to right). Then type "d = ". The answer will appear if in Automatic Mode. Or press F9.

$$a := 1.0 \qquad b := 4.5 \qquad c := 6.0 \qquad \text{definition of variables}$$
$$e := 34.0 \qquad f := 9.0$$
$$d := \frac{(a+b)^2}{c} + \sqrt{e} \cdot f \qquad \text{definition of equation}$$
$$d = 57.52 \qquad \text{evaluation}$$

Figure 3

By moving the cursor into the respective definition region, each of the variables can be changed.

As you move the cursor out of the newly edited region, you should see the answer (d =) alter in response to your changes. This automatic recalculation is similar to the behavior of a spreadsheet program. If the Automatic Mode is switched off, press F9.

Range of Variables

If one of the variables within the equation has a well defined range over which it varies, then Mathcad allows a succession of inputs to the same expression. If the variable *e* takes on integer values from 34 to 40, the output *d* must be defined in such a way to accept multiple inputs and output them sequentially.

The range specifier for *e* is defined by typing " ; " (semi-colon, without the quotes). Edit your expression for *e* by moving the cursor into the region and clicking near the number 34. Select the number using the up-arrow or spacebar. A selection box should appear with its upper-right corner notched. Now type " ; " and add the number 40 to the end of the description.

If you are in the Manual mode, press F9 or select Math/Calculate. An error message (non-scalar value) will appear in the region of the definition of the equation. Although you have allowed *e* to vary, you have not allowed a similar flexibility in d. By selecting *d* from its expression, add (*e*) to it so that the left-hand side of the equation now reads *d(e)=*. Press F9 again and a table of output values should appear. The values of *e* can likewise be displayed by typing " *e* = ". Figure 4 shows the output for these changes.

$a := 1 \qquad b := 4.5 \qquad\qquad c := 6 \qquad\qquad e := 34 .. 40 \qquad f := 9$

$$d(e) := \frac{(a+b)^2}{c} + \sqrt{e \cdot f}$$

Here is the output for the function d(e) with the input varying over a succession of integers

e	d (e)
34	57.52
35	58.286
36	59.042
37	59.787
38	60.521
39	61.247
40	61.963

Figure 4

You have created a function *d(e)* and allowed the input to drive the output as the other variables stayed fixed.

Save this file under a different filename.

Maple

The equation for *d* in the first exercise is static and numeric. The variables represent numbers which have to be defined. Otherwise an error message will result. And, if you select one of the variables to evaluate, the number associated with it will be returned on the other side of the equal sign.

To re-express the equation for one of the other variables, a reorganization of the symbols is needed. Mathcad accomplishes this by incorporating a subset of Maple, a powerful symbolic manipulator from Waterloo Maple Software. In Mathcad 5.0 select Symbolic/Load Symbolic Processor to load the Maple subset. A maple leaf icon will appear. In Mathcad 6.0, the Symbolic Processor (now unidentified) is automatically loaded upon start-up.

The small portion of the full Maple program which is included in Mathcad complements it well enough.

Rewrite the original expression for the equation *d* using CTRL-= in the place of the definition symbol. Select one of the variables on the right-hand side of the definition symbol.

☑ Selection is accomplished by moving the cursor into the equation region and clicking. If the notched box contains more than you need, de-select by pressing the down arrow key until a single vertical line is seen. Move the line using the right and left arrow keys and, when the line is to the immediate left of the desired symbol, press the up-arrow key. The notched rectangle should now enclose the variable.

Once the variable has been selected, choose Symbolic/Solve for Variable. The new equation(s) should appear immediately to the right or below the original expressions depending on the derivation preferences you have selected. Figure 5 contains the original expression of *d* and the re-expression of the statement for the variable *b*.

$$d = \frac{(a+b)^2}{c} + \sqrt{e \cdot f} \qquad \text{has solution(s)} \qquad \begin{bmatrix} \frac{-1}{2} \cdot c \cdot \left(\frac{2}{c} \cdot a + 2 \cdot \frac{\sqrt{d} - \sqrt{e \cdot f}}{\sqrt{c}} \right) \\ \frac{-1}{2} \cdot c \cdot \left(\frac{2}{c} \cdot a - 2 \cdot \frac{\sqrt{d} - \sqrt{e \cdot f}}{\sqrt{c}} \right) \end{bmatrix} \qquad \text{for } b$$

Figure 5

Try this process for other selections of the variable. Here, you have the opportunity of re-expressing the original mathematical sentence in a variety of ways.

Although each new sentence contains the same relations between the variables, the subject of the sentence changes relative to the verb "is".

Explorations

1. Determine the value of *a* in the equation $a = \frac{(b+c)^2}{d}$ for $b = 2$, $c = 3$ and $d = 4$. First define *b, c* and *d* and the equation for *a*. Definition merely associates a constant with a variable or a set of variables within an equation to another variable. To determine the output, the equal sign (=) forces the calculation.

2. Determine the value of *a* in number 1 above for each of the following sets of *b, c* and *d* values:

   a) $b = -3.4$, $c = 15.6$, $d = 9.2$
   b) $b = 2.456$, $c = -3.654$, $d = 7.825$
   c) $b = 0.5$, $c = -0.245$, $d = 0.0001$
   d) $b = 10^{-4}$, $c = 3.2 \times 10^{-1}$, $d = \pi$

   The symbol π is typed using the key combination CTRL-p. The exponents are written by pressing the ^ (caret) key.

3. For the function, $f(x) := x^3 + x^2$, create data tables for the independent and dependent variables over the domain [-5, 5] in unit steps. The exponents are created by typing the ^ (caret) symbol. As you type, the exponent moves into place as a superscript. Save

the file by pressing F6 and providing a suitable filename. Unless you have specified otherwise in the 'Working Directory' field of the File/Properties... dialog box from the Program Manager, the file will be saved to your Mathcad directory. You may instead want to create a separate file area for your data files so as to keep program and data isolated. Backup of your data files is easier if they are all stored in one convenient location. 2. Edit the function in #1 to $f(x) := 3x^3 + 2x^2 - 4$ over the expanded domain [-10, +10]. Examine the effect of choosing step-sizes of 1, 0.1, 0.01 and 0.001 on the design and effectiveness of the worksheet.

4. Examine the function $f(x) := ax^3 + bx^2 + cx + d$ for assigned values of a, b, c and d. Generate tables of values for x and $f(x)$ over a suitable range and with an appropriate step-size. Examine the effects of altering any defined variable or constant.

5. Experiment with the creation of an equation of many variables. Load the Symbolic Processor. Use the Symbolic/Solve for Variable option to express the equation in as many ways as there are variables (isolate one variable at a time by selecting it from the equation). As a sample, try $V = I \cdot \dfrac{Rr}{R+r}$.

6. Use the Symbolic/Solve for Variable option to isolate the following variables:

    a) $V = I(R + r)$ for r

    b) $PV = nRT$ for R

    c) $R \cdot A = \dfrac{R \cdot r}{R + r} + B$ for R

    d) $L = l + \alpha l(\Delta T)$ for α

The Δ (delta) symbol in the last equation is typed by pressing the CTRL-g key combination after typing D.

7. Remember to Save, Save, Save. The simple press of the F6 button or a click on the disk icon every two or three minutes can prevent you from losing hours of work. Experience shows that everyone who uses a computer must survive the loss of their work at least once. Try to avoid the heartbreak.

1.2 The Handbooks

All the Windows versions of Mathcad (versions 3.0 to 6.0) have a reference feature called Handbooks. These Handbooks either come packaged with the version you are using or can be bought separately depending on your interest. The Handbooks cover: General Scientific, Desktop Reference, Electrical Engineering, Applied Mathematics (differential equations), and much more.

Mathsoft Inc. once published 'Applied Mathcad', a quarterly electronic magazine. The magazines were full of extremely interesting technical applications, historical examples and demonstrations of mathematical concepts. Each magazine was stored as a Handbook and accessed through the Books/Open Book option from the menu.

However, there is no longer a need for this snail-mail disk system as all of the Mathcad sample and research files can be accessed from the Mathsoft World-Wide-Web site at http://www.mathsoft.com. And, after you have gained a command of the software, you may wish to upload one of your particularly nice solutions to the Web site to share with others.

The real strength of these online references is their hypertext structure (as a Help file should be) and their transportability into the current working documents. Text, graphics, pictures, and equations can be copied from the Handbook and pasted directly into your working document. If the region is a mathematics region, it is brought in live. As long as all constants and variables have been defined, the newly imported equation becomes part of the calculation. And, if need be, these equations can be edited.

The process of copying and importing does not affect the reference Handbook. Although you can place bookmarks within the Handbook, you have no access to editing the content. So, the next time you need a particular piece of information, you can be assured it will be available.

Warmup

As an example of the use of the Handbooks, open the Desktop Reference Handbook by selecting Books/Desktop Reference from the menu at the top of the working document. The Handbook is formatted so that a worksheet window may be tiled with the Handbook window with a minimum of lost information.

Another valuable Book, Getting Started, contains full samples of Mathcad documents and is a welcome resource if you are just starting out.

The Handbook will open on its cover and
- in Version 5.0, the vertical icon bar will be replaced with an EB (electronic book) icon list.
- in Version 6.0, a row of icons will appear above the Handbook region.

The two most important of these icons are TOC (table of contents) and Index. The other icons let you flip from page to page or from section to section.

Within the TOC of the Desktop Reference, scroll to Section 3 on Geometric Formulas. Double-click (select) on the text region of section 3.2, Volumes and Surface Areas. You should now find yourself in another menu area. Scroll down the page until you see the picture of the Sphere. By double-clicking on the text to the immediate right of the picture, you are brought into the relevant Handbook region.

Another way to find a topic is to select the Books/Search Books... option. By typing in the words Volume or Sphere, any existing matches are displayed.

Figure 1 shows the Handbook region for the volume and surface area of a sphere. By selecting the picture, text, and mathematical regions, we can copy them into our working document by using:
- the copy, cut and paste icons,
- the menu options (Edit/Cut and Edit/Paste), or
- the keyboard shortcuts (CTRL-C, CTRL-X and CTRL-V).

The Handbook section also contains a fully worked out example. This example can be imported into your document like any other area of the Handbook. However, in some instances, the time it takes to find and to cut and paste the relevant section ends up exceeding the time it would have taken to simply write your own definitions. If the use of the Handbooks is lengthening the process, you may decide to leave them on their virtual shelves.

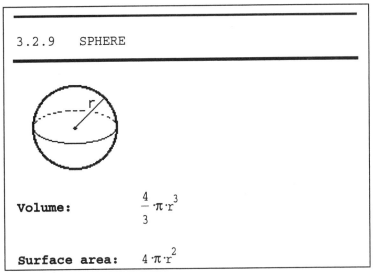

Figure 1

Figure 2 shows the picture, text and mathematical areas copied into a Mathcad document and rearranged.

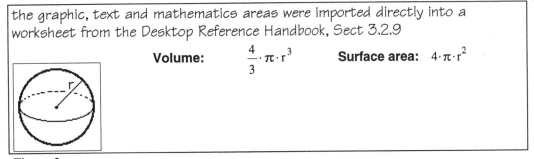

Figure 2

At this point the relations are not mathematically active although they are now part of an active worksheet. The radius, volume and surface area relations have not been properly defined.

In Figure 3, the mathematical expressions for the volume and surface area of a sphere have been copied into their respective definitions. A radius has been defined and units have been included.

The centimeter unit (cm) was included within the radius definition by multiplying the number of centimeters by the expression 'cm'. Mathcad includes a number of pre-defined units including 'cm' and its prefix variants (m, mm, km, etc.). Section 1.1 of the Desktop Reference Handbook discusses the use of units in detail.

The calculation of the volume and surface area is invoked by using the '=' sign applied to the variable name. A place holder ■ appears to the right of the equation region and can be replaced by any predefined and consistent unit. Mathcad will automatically convert between unit systems.

Note that Mathcad has taken care of the unit conversions for you and has displayed the answers to the default level of accuracy (to 3 decimal places and an exponential threshold of 10^3 or 10^{-3}).

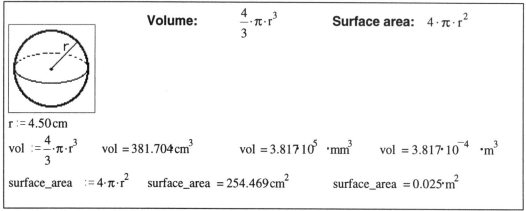

$$\text{Volume:} \quad \frac{4}{3}\cdot\pi\cdot r^3 \qquad\qquad \text{Surface area:} \quad 4\cdot\pi\cdot r^2$$

$$r := 4.50\,cm$$

$$vol := \frac{4}{3}\cdot\pi\cdot r^3 \qquad vol = 381.704\,cm^3 \qquad vol = 3.817\cdot10^5\;\cdot mm^3 \qquad vol = 3.817\cdot10^{-4}\;\cdot m^3$$

$$surface\_area := 4\cdot\pi\cdot r^2 \qquad surface\_area = 254.469\,cm^2 \qquad surface\_area = 0.025\cdot m^2$$

Figure 3

Explorations

1. Use the Desktop Reference Handbook to find the surface area and volume of a frustum of a right circular cone (a cone with its point chopped off parallel to the base). The radius of the base is 8.0 cm, the height is 4.5 cm and the radius of the topmost circular area is 5.0 cm. Try to express your answers in a variety of units.
2. Choose a section of the Desktop Reference Handbook and edit the content (text or mathematics). From the Books menu, select Annotate Book. Any changes you make will be preserved for reference. The original copy, however, remains untouched. To return to the original, deselect Annotate Book.
3. Use the Desktop Reference Handbook to find the surface area and volume of a spherical cap section. The original sphere has a radius 6.0 m. The cap's base (an intersection of the sphere and a plane) is 3.5 m from the center of the sphere along a radius.
4. Determine the capacitance of a parallel plate capacitor with an area of 1.0 cm^2 and an inter-plate distance of 0.0001 mm. In the Desktop Reference, the default permittivity ε is that of the air between the plates. Determine the change in capacitance if, instead, rubber fills the space between the two plates. (This exercise assumes no knowledge of capacitors or electromagnetics. All of the information can be found by searching through the Handbook).

1.3 Formatting: Numbers, Graphs, Text

In this set of explorations, we will be examining the look of the documents created in Mathcad. The alteration of the style of the document may not seem very important at first; after all, the content of the document is what you are attempting to communicate. Your elegant solution to an existing problem or your creative approach to the understanding of a complex idea through exploration is your primary concern.

Mathematics is a language, a formalized attempt at communication. It possesses a particular vocabulary, grammar and syntax. Although it may appear slightly more rigid in its rules than other languages, the possibility of developing personal expression still exists.

An attribute usually not associated with mathematics is style. Mathcad offers a rich environment for developing a personal style. No two people will solve a problem in the same way. Just as no two people express themselves the same way. In this manner, self-expression is more clearly visible.

Your solution or exploration will be equally correct whether it is displayed in one style or another. The ability to alter the look of your communication will not alter the content. However, clarity may be gained by using some of the tools examined in this set. Your ability to format the various types of information will allow a clearer numerical, graphical or text-based description of the exploration at hand.

And, as with all highly editable environments, you have to know when to stop.

Warmup

Numbers

Mathcad offers a variety of options under the heading of numerical formatting. If you create a table of values for the function $f(x) = x^2$ over the domain [-2, +2], Mathcad uses its default settings to represent the table of values for x and $f(x)$. In Figure 1, the first two columns represent the output of the domain and range for these default settings. There are no trailing zeros and the displayed precision (number of decimal places, here represented as dp) is set to 3. By double-clicking anywhere within the table, a Numerical Format dialog box appears. The third table of Figure 1 shows the effect of turning the Trailing Zeros option on while the fourth table adds two more decimal places to the Displayed Precision (3) for a total of 5 decimal places. The numbers included within brackets are the default settings.

Of particular importance in the look of your numbers is the set of complementary radio buttons at the bottom of the Numerical Format box. With these, you can select whether your particular settings apply to all numbers within the document or only to those within the table (or calculation area) you have selected. If you wish to apply your preferences to all numbers, choose GLOBAL. If you are applying the formatting options to a more restricted area, choose LOCAL. Keep in mind that a LOCAL format will overtake any GLOBAL formatting you may have previously selected for the document.

Your head may already be swirling from the wealth of choices. The unfamiliar becomes familiar through use. Start with the default settings, follow and extend the suggestions in the Warmup. Play with each option until you are comfortable with the effected change in the output. And remember, whatever you do to the 'exterior' look of the numbers, Mathcad is still calculating them to its maximum 'internal' accuracy level of 15 decimal places.

$x := -2 .. 2$ $f(x) := x^2$

x	$f(x)$	$f(x)$	$f(x)$
-2	4	4.000	4.00000
-1	1	1.000	1.00000
0	0	0	0
1	1	1.000	1.00000
2	4	4.000	4.00000
	default	zero on dp at 3	zero on dp at 5

Figure 1

The effect of increased precision on the look of the numbers is more clearly displayed in Figure 2. Here, the function $g(x) = \sqrt{x}$ has been represented. The first output table for the function (column 2) displays the default settings of 3 decimal place precision and trailing zeros turned off. Column 3 indicates the trailing zeros turned on while the next columns, with trailing zeros off, indicate displayed precisions of 5, 7 and 15 decimal places respectively.

$x := 0, 0.5 .. 3$ $g(x) := \sqrt{x}$

x	$g(x)$	$g(x)$	$g(x)$	$g(x)$	$g(x)$
0	0	0	0	0	0
0.5	0.707	0.707	0.70711	0.7071068	0.707106781186548
1	1	1.000	1	1	1
1.5	1.225	1.225	1.22474	1.2247449	1.224744871391589
2	1.414	1.414	1.41421	1.4142136	1.414213562373095
2.5	1.581	1.581	1.58114	1.5811388	1.58113883008419
3	1.732	1.732	1.73205	1.7320508	1.732050807568877
	def	zero on	zero off dp at 5	dp at 7	dp at 15

Figure 2

Within the selections, you also have control over the exponential threshold, the point at which the numbers are displayed in exponential notation. A setting of 4 would force the output of any number whose absolute value is greater than 10^4 or less than 10^{-4} to be represented in exponential notation. As an example of the need for this option, a result of 0.0001 with Displayed Precision set at 3 while the Exponential Threshold is set at 5 would return an output of 0 (zero).

Figure 3 shows output tables for the function $h(x) = 10^9$. The Exponential Threshold has been set to 3, 5 and 7 in the indicated tables. For the right-most table, the exponential threshold has been set to its lowest value of 1 while the trailing zeros have been turned on and the displayed precision set at 3. This setting is equivalent to scientific notation to 3 decimal place accuracy.

☑ In Figures 1 through 3, the font (Times New Roman) associated with variable names has been italicized. The cursor was placed beside the variable name on the worksheet and the Italic button was selected from the Font Bar. Although variable and function *names* are usually italicized, Mathcad also italicizes the associated *numbers*. So to preserve numbers in a regular form, variables within the Mathcad documents represented in all other figures in the book have not been italicized.

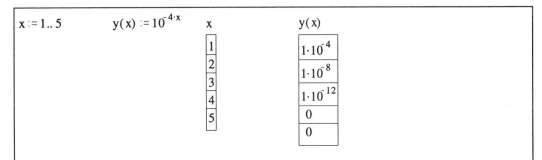

Figure 3 content:

$x := 0, 0.5 .. 3$ $\qquad h(x) := 10 \cdot x^9$

x	h(x)	h(x)	h(x)	h(x)
0	0	0	0	0
0.5	0.02	0.02	0.02	$1.953 \cdot 10^{-2}$
1	10	10	10	$1.000 \cdot 10$
1.5	384.434	384.434	384.434	$3.844 \cdot 10^2$
2	$5.12 \cdot 10^3$	5120	5120	$5.120 \cdot 10^3$
2.5	$3.815 \cdot 10^4$	38146.973	38146.973	$3.815 \cdot 10^4$
3	$1.968 \cdot 10^5$	$1.968 \cdot 10^5$	196830	$1.968 \cdot 10^5$
	exp 3	exp 5	exp 7	scientific notation

Figure 3

Two other options within the Numerical Format dialog box are of present interest.

If you locally format a set of numbers, notice the Zero Tolerance option will be grayed out. This GLOBAL option sets the limits beyond which Mathcad treats numbers as zeros. And, counter to the usual over-riding of GLOBAL formats by LOCAL formats, this option cannot be locally set.

The default setting for this option is 15. Any number whose absolute value falls below 10^{-15} is treated as a zero.

Figure 4 shows the function $y(x) = 10^{-4x}$ as locally formatted. As the output grows beyond the limit, zeros are output. This limit can be adjusted from 0 to 307.

$x := 1 .. 5$ $\qquad y(x) := 10^{-4 \cdot x}$

x	y(x)
1	$1 \cdot 10^{-4}$
2	$1 \cdot 10^{-8}$
3	$1 \cdot 10^{-12}$
4	0
5	0

The LOCAL radio button is turned on. The zero tolerance is set to the default of 15. Notice that numbers less than 10^{-15} are interpreted as zeros. Changes in the Zero Tolerance adjustment are not applicable to LOCAL settings.

Figure 4

Figure 5 shows a GLOBAL format setting for the Zero Tolerance of 5. This setting would apply to all numbers throughout the document and cannot be overwritten by the LOCAL setting.

$x := 1 .. 4$ $y(x) := 10^{-2 \cdot x}$

x	y(x)
1	0.01
2	$1 \cdot 10^{-4}$
3	0
4	0

the GLOBAL radio button has been turned on and the Zero Tolerance set to 5. Notice that numbers less than 10^{-5} are interpreted as zeros

Figure 5

The final option within numerical formatting is the Radix. We will examine the selection of complex number representation in a later set of Explorations.

The Radix defines the base of the numbering system used to represent numbers. Although the decimal system is the most familiar to us and is always as far away as our ten fingers, it is by no means the only numbering system or even the best numbering system depending on the situation.

In the world of binary arithmetic, where the base is 2 and the numbers require many digits for their expression, more compact forms of the numbers are used. The Octal system, base 8 or 2^3, can be used to represent a binary number using one character for every 3 characters used in Binary. And, similarly, the Hexadecimal system based on powers of 16 or 2^4 decreases the ratio to 1:4.

Figure 6 displays positive and negative integers in the three forms of decimal, octal and hex(adecimal). The last set of tables attempts to represent decimal fractions in all three forms. As can be seen from the tables, only integers are translated from one form to another.

$x := 20, 40 .. 100$ decimal, octal and hexadecimal forms

x	x	x		- x	- x	- x
20	24o	14h		- 20	- 24o	- 14h
40	50o	28h		- 40	- 50o	- 28h
60	74o	3ch		- 60	- 74o	- 3ch
80	120o	50h		- 80	- 120o	- 50h
100	144o	64h		- 100	- 144o	- 64h

positive integers   negative integers

$x := 1.0, 1.2 .. 2.0$ decimal, octal and hex formats for decimal fractions

x	x	x
1.0	1o	1h
1.2	1o	1h
1.4	1o	1h
1.6	1o	1h
1.8	1o	1h
2.0	1o	1h

Figure 6

Graphs

If you thought the formatting of numbers has a bewildering array of options available, the wealth of different ways of representing graphical information may surprise you. Here, we will concentrate on two-dimensional plots of simple functions. Mathcad allows far more complex representations of three-dimensional information and, in its Mathcad Version 6 Plus format, allows you to animate your graphics.

A confused and overly messy picture is worth far fewer than a thousand words. If the graphic is cluttered, it presents the same confusion to the eye as endless tables of data accurate to 15 decimal places. There is simply too much information presented. The reader then either spends a ridiculous amount of time deciphering your information or, in most cases, simply skips over it no matter how much work has gone into its presentation.

We will examine a simple parabolic function, $y(x) = x^2$ over the domain [-3, +3] and represent it in as many ways as possible. Some of these may touch on the side of confusion, making the plot more work than it has to be. Ultimately, you have to decide on the look of your plots.

To create a simple plot region, first define the range of the input variable, its step size and the function itself. An example of this process was covered in Section 1.1. To define a rectangular plot region, click on the graphics icon from the Symbol Palette. This will bring up a secondary palette with various plot types indicated. Click on the rectangular grid pattern in the top left corner. (Note: this Palette is available in Version 6 only). Otherwise, the rectangular plot region may be accessed by pressing the @ key or by selecting Graphics/Create X-Y Plot from the pull-down menus.

An empty plot region will appear with 6 empty placeholders. Figure 7 shows what you should expect to see after defining the function, domain and after having created a plot region. At this point, if you try to represent the function, an error message ("missing operand") will appear. You need to define the input and output variables along the horizontal and vertical axes.

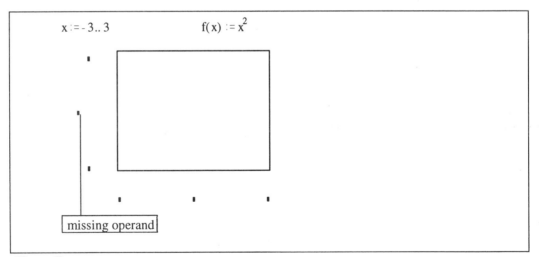

Figure 7

Click on the plot region to enter the editing mode. To begin with, add the *x* and f(*x*) to the empty placeholders in the middle of the horizontal and vertical axes respectively. Movement to the selection area can be accomplished by either using the mouse and clicking on the empty placeholder or by repeatedly pressing the TAB button until the notched rectangle lands on the desired placeholder.

If you are in the Automatic mode, the plot should appear as soon as you click outside the plot region. Otherwise, press F9. The plot you have created should resemble that in Figure 8 (a).

With a step size set to 1, the plot is represented by straight line connections between the 7 coordinate pairs. As you can see in Figure 8 (a) the plot is jagged. Even though the trend of the curve is obvious, its quadratic nature is obscured by the lack of detail. Figure 8 (b) is an attempt to remedy the problem by increasing the number of points used. The quadratic becomes smooth and well-behaved (no sharp points).

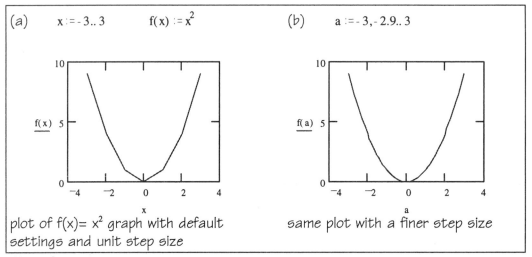

(a) $x := -3 .. 3$ $f(x) := x^2$ (b) $a := -3, -2.9 .. 3$

plot of f(x)= x² graph with default settings and unit step size

same plot with a finer step size

Figure 8

Note as well that the other information has been automatically included. Scale marks and numbers have been added to help clarify the information.

The plot domain extends only from -3 to +3. The function's representation ends at these limits. If we wish to extend the domain, we can simply edit the numbers within our definition of *x*. However, if we would rather keep the same domain but change our viewing scale, we can edit the end points of the axes.

Select the plot region. Four numbers will appear which represent the axis limits. Click on the *x*-axis limits and change them to any other number. In Figure 9, the limits have been changed from -4 and +4 to -2 and +2 respectively. Although the function's range has not been changed, the effect is to change the vantage point of the viewer. You have moved into the function and now see it from a little closer.

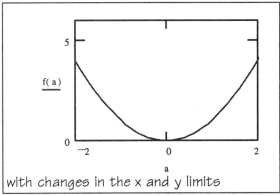

with changes in the x and y limits

Figure 9

Figure 10 shows the result of repeated changes to the same plot. These changes were created by double-clicking on the selected plot region and then making selections from the many options presented. In Version 6, the options X-Y Axes, Traces, Labels and Defaults are presented on a series of tabbed folders. In earlier versions, all of these options were presented within a single box. Below each figure is a description of the location of the option and the change.

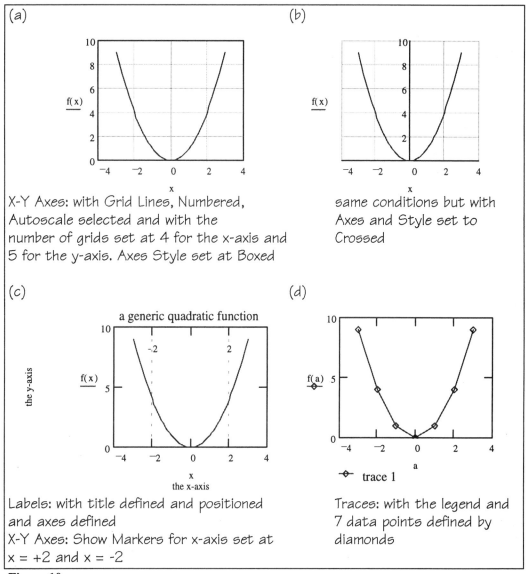

Figure 10

Of course, the option list does not end with Figure 10. The plot can also be represented in bar chart and step function formats (Figures 11 (a) and (b)). The plotted points themselves can also be highlighted by separate symbols. This last feature should be used wisely as the plots in Figures 11 (c) and (d) show.

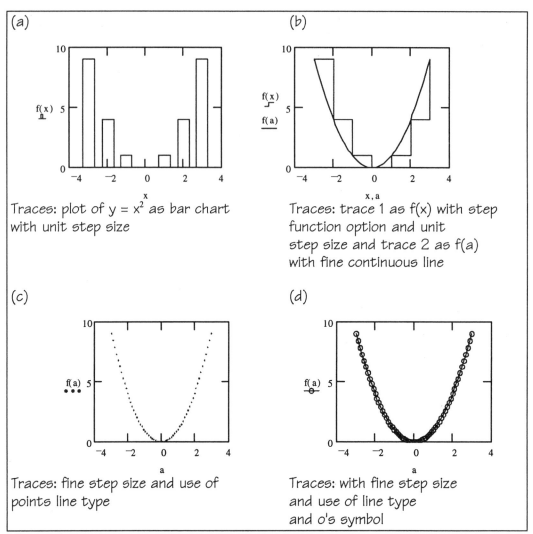

Figure 11

The trace width (Figure 12) can also be adjusted for clearer printed and viewing display by altering the Traces, Weight option. This feature was not available in previous versions of Mathcad. And, while the feature is not that useful for display on a color monitor, printed plots and plots projected through an LCD panel can sometimes appear as faint traces only for the default weight of 1.

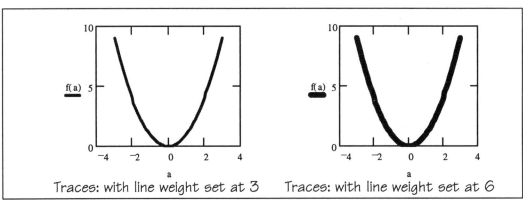

Figure 12

As the last (but by no means final) option, the same function may be represented in different plot formats selected from the graphing palette. The secondary graphing palette contains a polar plot representation in the upper right-hand corner. This format can also be accessed by selecting Graphics/Create Polar Plot or by using the key combination CTRL-7. Figure 13 shows our familiar quadratic seen as a polar plot. Within the polar plot format, the input variable x represents an angle in radians measured counter-clockwise from the direction specified by the zero. The function $f(x)$ represents the distance the point is from the center along the angle specified by x.

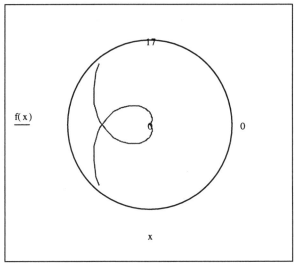

Figure 13

A more symmetric polar pattern is given in Figure 14. As x increases from 0 to 2π, the radial distance $r(x)$ is controlled by the cosine function. The jaggedness of the plot could be smoothed by increasing the number of points over the domain.

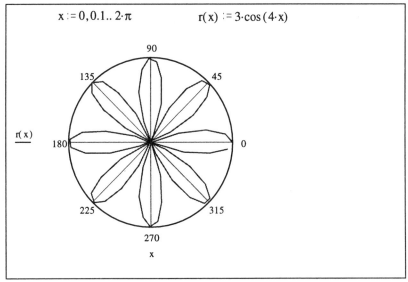

Figure 14

Text

Although you are communicating mathematics in these explorations and probably dismiss the need for words to describe your solutions, the majority of the learning takes place when you become aware of the process you have chosen to execute.

Most of my students felt that they learned the mathematics when they had to communicate their ideas in words. And, the use of text to describe your steps is a worthwhile habit to develop if you are thinking of pursuing a programming career. Words help define the methods used.

Descriptive text is inserted into a Mathcad document by using the double quotes, ". Or you can select Text/Create Text Region from the menu bar. This creates a text region at the cursor position which can be formatted for font type and size, for characteristics such as bold, italic or underlined text, for sub- and super-scripts and for text region width.

As a Windows program, Mathcad offers a variety of True Type Fonts (TTF) available to your document. These selections are applicable to the variables and constants within the mathematics regions and to the text in the non-active regions.

The default selections will appear in the Font Bar just above the worksheet space on your screen.

For each of these types of information, there may be specific fonts particular to the screen (*S) or printer (*P). These are probably best avoided so that you can take advantage of the visual environment offered by Windows. And, rather than overwhelm the senses of your reader with a wealth of varying fonts, sizes and highlighting features, aim for clarity. The consistent use of a few easy-to-read fonts will help out rather than confuse the process of understanding.

As an example of the font tags applied to the mathematically active region, Figure 15 shows the variable *x* represented in Times New Roman and its associated constants in Ultra Shadow. The overall result is less than successful.

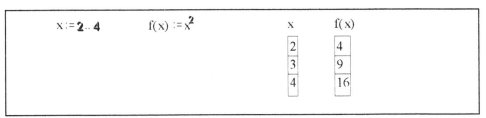

Figure 15

Figures 16 and 17 illustrate some of the options available within the descriptive text regions.

Here is an example of my default selection for text, Technical font, size 12.
Here is an example of a selection for text, Arial font, size 12.
Here is an example of a default selection for text, Technical font, size 14.
Or you can choose to highlight text in BOLD, *Italic*, or <u>underline </u>by using the **B**, *I* or <u>u</u> buttons within the Font Bar.

Figure 16

Subscripts (x_2, a_b) and superscripts (x^2, a^b) can be added as needed using the a_b and a^b buttons on the Font Bar.

Here is an example of the width of the text region set by using the key combination CTRL-Enter. The width of the region may be increased by selecting the area with the dashed selection rectangle and dragging the right vertical line.

Figure 17

With Version 6.0, equations can be inserted into the text regions so that the distinction between the mathematically active region and the mathematically inert text begins to blur. To insert an equation in a text region, place the cursor at the point of insertion and select Text/Embed Math from the menu. Insert the equation just as you would outside a text region.

The equation is active. There is no reason to redefine it outside the text region if that is the effect you are after. However, if the equations are for descriptive purposes only, they may be made inactive by selecting the equation (clicking on it) and then selecting Math/Toggle Equation from the menu. A small rectangle will appear to the immediate right of the equation as a reminder of its having been turned off. Figure 18 illustrates an inactive formula included within descriptive text.

The equation for the one-dimensional (vertical) position of an object of mass 'm' with initial position x_0 and initial velocity v_0 in a gravitational field 'g' is given by $x(t) := 0.5 \cdot g \cdot t^2 + v_0 \cdot t + x_0$. Here the equation has been toggled off.

Figure 18

The Configuration File

At some point in your development of the explorations (your own or those suggested within this text) you may want to save the look of your page. Your choice of fonts and fontsizes for text, variables and constants, the style of your plot types may become an expression of your particular style. In this case, you can save the preferred environment by selecting File/Save Configuration. A file, MCAD.MCC, will be created or re-written in the root Mathcad directory.

You may however have different preferences for different solution strategies. In this case, you may save a custom configuration file under a different filename with the extension .MCC and, after selecting File/Execute Configuration File, change the look of your document at anytime within your Mathcad session.

At this point, explore the variety of configurations and, when you have exhausted yourself, save the screen settings as an *.MCC file.

Conclusion

The strength of any highly editable medium is its capacity for self-expression. This, however, is also its downfall. Experience with these tools will allow you to discover your preferred style. Use the feedback you receive from colleagues to refine your worksheets to the point where the content is communicated clearly within a uniquely personal fashion.

Explorations

1. Create a default representation of the plot of $y(x) = x^3$ from $x = -3$ to $+3$. By using various plot types, line thicknesses, markers, titles, labels, colors and whatever else you can click on, create 10 distinctly different styles of the same plot information. The plots can all be saved in one large file or separately, in ten smaller ones. You may find the former option allows easier comparison while the latter allows quicker loading and processing times.

2. Now add descriptive text regions which explain, in detail, the changes you have created for each of your ten plots.

3. For the function $f(x) = \sin(x)$ over the domain [0, 2], generate tables of values for x and $f(x)$ with a step size of 0.1. Select the table in $f(x)$ and copy it a number of times using either the Copy and Paste icons or the keyboard shortcuts (CTRL-C, CTRL-V). For each copied table, double-click on the table region to access the Numerical Format dialog box. Then choose different values of the displayed precision and exponential threshold. Document your choices so that you may refer to them later.

4. The easiest way to gain insight into formatting is to examine how others have formatted their work. Open files within the drive:\winmcad\handbook\getstart directory. These *.mcd files are all highly editable. Change the fonts used by the variables and constants. Reorganize text into paragraphs of well defined width. Alter the look of the graphs. And save your new creation with a different filename. There are also many *.mcd files in the directory drive:\winmcad\qsheets, the Quicksheets tool available in Version 6.0.

5. Open one of the sample Mathcad files (or one of your saved files) and alter the design of the document by choosing other font types and sizes for the variables and constants. Don't worry about the content at this stage. By clicking near a region and dragging the mouse, a dashed-line selection window should appear. Release the mouse. A bold plus-sign now indicates that you can now drag the section anywhere on the workpage. Experiment with dragging and dropping text, tables, plots. Or you can copy, cut and paste any selected region using the pull-down menus or keyboard shortcuts. Significantly alter the look of the original file and save it under another name (e.g.: new-look1.mcd).

6. For the file which you have created in #1, save the configuration by selecting File/Save Configuration file. If you wish to make this style the default, choose the 'mcad.mcc' filename. Otherwise, select another filename with the '.mcc' file extension.

Chapter 2: Basic Algebra

Algebra's great power is its ability to generalize the specific.

In an applied sense, specific numerical observations lead to the discovery of an underlying pattern within a system. Instead of having to refer to tables upon tables of prerecorded input-output data every time a new bit of input data emerges within the system, the derived algebraic relation allows a quick and ready tool into which the new data can be input and the new result output. If the algebraic relation is useful, then it should not only produce output which can be tested against past experiences, it should also allow predictive output.

Thus, the problem need not be re-solved every time a variable number changes.

The techniques of algebraic manipulation are often presented as the end of algebra. However, here, the point is to solve problems and explore. With that view in mind, this chapter will concentrate on the use of Mathcad's built-in symbolic manipulator. We will examine its use so that the technique of algebraic manipulation becomes trivial and attention can then be paid to solving the problems at hand.

Computer algebra systems (CAS) seem to draw a host of negative reactions from those for whom the algebraic end is the algebraic technique. And there are valid criticisms. If the understanding of the underlying routine is overlooked, then indeed the results seem to appear mysteriously from thin air. And, the user will have no way of validating the responses since they have no direct experience of the process.

However, in our case, try to approach the symbolic tools as steps within a larger solution strategy and not as the end in themselves.

The original strength of Mathcad was in its use as a mathematical scratchpad, a useful tool used in the initial stages of the solution of complex problems. As it stands now, Mathcad's definition has changed to that of a robust development tool, a blend of strong numerical and symbolic tools.

There are limits to all this symbolic evaluation. High degree polynomials may not have symbolic forms for the definition of their roots. Other equations simply have no closed-form solution. You may then have to apply visual tools (graphs) and numerical tools to find the solution(s).

By blending algebraic, graphical and numerical solution strategies, you have not only the best chance of solving a problem completely but also of understanding it completely.

2.1 Algebraic Expressions and Maple

In its earliest incarnations, Mathcad was solely a numerical processor. All calculations were performed as floating-point numbers. Mathcad could not recognize the manipulation of symbols unless they had first been identified as numbers. Although Mathcad was still a powerful and useful tool, the algebraic tools of simplification and factoring and the exact-form integration and differentiation tools of calculus were not available to the user.

As an example, the derivative (a measure of the rate of change) of a function could be generated numerically and then compared to its exact analytic form as long as the user could determine that exact form outside the Mathcad environment.

With the inclusion of the symbolic processor these limitations have been overcome. Expressions can be simplified, factored or expanded. Exact symbolic forms of derivatives and integrals can be generated. As with all other elements within the worksheet, these elements are editable. They can then be evaluated numerically or be included in some other numerical process. If a floating point result is required, a simple click provides it.

In Version 5.0 of Mathcad, the symbolic processor must be loaded after the basic program has been loaded. Select Symbolic/Load Symbolic Processor from the Toolbar. During the loading process, a maple leaf symbol will appear on the screen identifying the processor as a subset of Maple, the symbolic mathematics package developed by Waterloo Maple Inc.

In Version 6.0, the symbolic processor is loaded at startup along with the rest of Mathcad. And there is no mention of Maple. Although the loading time is increased, the symbolic processor is always ready at hand.

In this set of explorations, we will examine the use of the symbolic processor as applied to basic algebra. The main hurdle is the use of the notched selection rectangle in the correct identification of the selected variable or expression.

Warmup

We will examine expansion, simplification by the collection of like terms, substitution of variables and expressions, the arithmetic of algebraic expressions and the formatting of the displayed results.

☑ Throughout these exercises, you do not need to define the variables used to be any particular numbers. After all, this lack of precise definition is the whole point of algebra. You may, however, run into conflicts within larger documents if the variables you are manipulating symbolically have already been defined numerically within the worksheet.

Simplification and Formats

After loading the Symbolic Processor in Mathcad Version 5.0 or loading Mathcad Version 6.0, move the cursor to a comfortable and empty location on the worksheet (at this stage the whole worksheet should be empty).

Type the following expression: $3ax + 6x^2 + 4ax + 5x^2$. As we need to refer to this expression later on within the Warmup, let's call it 'Expression One'.

☑ Even though this is a symbolic expression, make sure that you explicitly define all arithmetic operations. Mathcad will not interpret *3ax* as *3* times *a* times *x* unless you use the multiplication symbol (key: * or SHIFT - 8) between the variables and constants.

Now click anywhere within the expression. If you have clicked on an operator, a portion of the expression will be selected within a notched rectangle. If the click was closer to a constant or variable, a vertical line will appear.

Whatever happens, select the whole expression by
- repeatedly pressing the UP key, or
- repeatedly pressing the Spacebar.

At some point the whole expression will be selected and the repeated application of your trigger finger on the UP key will have no further effect. If you are using a trigger thumb on the Spacebar, repeated application beyond the selection of the full expression will de-select the expression and move the cursor to an empty position on the worksheet.

To simplify this expression by collecting like terms, select Symbolic/Simplify. Or, if you prefer the keyboard shortcut, use ALT-S then S.

Figure 1 shows the Simplify routine applied to the expression. Notice that the result appears immediately below the expression. You have some control over the format of this result and the inclusion of explanatory notes.

---

simplify process with no comments and with vertical orientation

$$3 \cdot a \cdot x + 6 \cdot x^2 + 4 \cdot a \cdot x + 5 \cdot x^2$$

$$7 \cdot a \cdot x + 11 \cdot x^2$$

Figure 1

Within the Symbolic pull-down menu, select Derivation Format (or ALT-S then o). The Derivation Format dialog box will appear. Select 'Show derivation comments' and select the radio button for 'horizontally'. Figure 2 shows the original simplification process with the derivation comments included. Figure 2(a) shows the derivation steps laid out horizontally while Figure 2(b) has the same comments arranged vertically.

---

a) comments on, horizontal

$3 \cdot a \cdot x + 6 \cdot x^2 + 4 \cdot a \cdot x + 5 \cdot x^2$ simplifies to $7 \cdot a \cdot x + 11 \cdot x^2$

b) comments on, vertical

$3 \cdot a \cdot x + 6 \cdot x^2 + 4 \cdot a \cdot x + 5 \cdot x^2$

simplifies to

$7 \cdot a \cdot x + 11 \cdot x^2$

Figure 2

If you are using the symbolic process as an intermediary (and erasable) step in a longer calculation process, the inclusion of explanatory notes and reminders may not seem that useful. However, as the stripped down process simply generates an answer alongside (or overtop) the original expression, you may find it useful to include derivation comments as a reminder of just what it is you are doing.

You may want the original expression to simply be replaced by its simplified variation. In that case, select Symbolic/Derive in Place. No comments will be shown and the simplified expression will appear in the worksheet space originally occupied by the longer expression. And, Derive in Place overrides any Derivation Format settings you have previously defined.

A variable within our original expression can be replaced by another variable or expression without having to go to the trouble of manually editing the expression one character at a time. To replace the variable x with the expression $(z + b)^2$, type $(z + b)^2$ somewhere beside 'Expression One'. Select the full expression in z and b and copy it to the Clipboard by selecting Edit/Copy (or CTRL-C). Return to 'Expression One' and select x by clicking beside it and using the UP arrow. Then select Symbolic/Substitute Expression. Not only has the variable been replaced everywhere with the replacement expression, the new total expression appears in simplified form. Very nice!

Figure 3 illustrates this substitution process.

$(z + b)^2$ replacement for x

$3 \cdot a \cdot x + 6 \cdot x^2 + 4 \cdot a \cdot x + 5 \cdot x^2$ original expression

by substitution, yields

$7 \cdot a \cdot (z + b)^2 + 11 \cdot (z + b)^4$

Figure 3

Expansion

The new expression can then be expanded using the distributive law by first selecting the expression then selecting Symbolic/Expand Expression (or ALT-S then x). Figure 4 shows the result of the repeated application of the Symbolic Processor as the original expression is transformed through substitution and expansion.

$(z + b)^2$ replacement for x

$3 \cdot a \cdot x + 6 \cdot x^2 + 4 \cdot a \cdot x + 5 \cdot x^2$ original expression

by substitution, yields

$7 \cdot a \cdot (z + b)^2 + 11 \cdot (z + b)^4$

expands to

$7 \cdot a \cdot z^2 + 14 \, a \cdot z \cdot b + 7 \cdot a \cdot b^2 + 11 \cdot z^4 + 44 \, z^3 \cdot b + 66 \, z^2 \cdot b^2 + 44 \, z \cdot b^3 + 11 \cdot b^4$

Figure 4

The substitution process also applies to any other variables within the expression. Figure 5 shows the replacement of the variable a in 'Expression One' with $(z + b)^2$. Again, the new expression has been simplified automatically.

$$(z + b)^2 \qquad\qquad \text{expression to replace a}$$

$$3 \cdot a \cdot x + 6 \cdot x^2 + 4 \cdot a \cdot x + 5 \cdot x^2$$
by substitution, yields

$$7 \cdot (z + b)^2 \cdot x + 11 \cdot x^2$$

Figure 5

The expansion process can be applied to any type of expression. Figure 6 illustrates a few examples.

(a)

$(x + 2) \cdot (x + 9)$
expands to

$x^2 + 11 \cdot x + 18$

(b)

$(a + b \cdot x)^4$
expands to

$a^4 + 4 \cdot a^3 \cdot b \cdot x + 6 \cdot a^2 \cdot b^2 \cdot x^2 + 4 \cdot a \cdot b^3 \cdot x^3 + b^4 \cdot x^4$

(c)

$2 \cdot a \cdot x \cdot \left(-4 \cdot x \cdot c^3 + 3 \cdot x \cdot c\right)$
expands to

$-8 \cdot a \cdot x^2 \cdot c^3 + 6 \cdot a \cdot x^2 \cdot c$

Figure 6

Multiplication and Division of Algebraic Expressions

The arithmetic of algebraic expressions includes the operations of multiplication and division. The Symbolic/Simplify and Symbolic/Expand have so far been applied to simple additive expressions.

The expression $\dfrac{x^3 y^{-3} m^6 n^2}{x^4 m^3 n}$ can be simplified using the Laws of Exponents. Mathcad applies these laws through the Simplify command applied to the whole selected expression. Figure 7 shows the original expression and its simplified version.

☑ For variables with exponents located within a product, make sure to select (using the UP arrow or spacebar) the variable and exponent before typing '*' for multiplication. When the whole of the numerator has been created, select it in total before pressing the ' / ' key for division. Then proceed with the definition of the denominator.

In Figure 7, values of *x, y, m* and *n* were defined after the region of the original and simplified expressions. Each expression was then copied to a blank space on the worksheet. The expression was selected with the selection rectangle's notch in the top-right corner. An equal sign was typed beside each of the selected expressions. This process showed that simplification does not change the numerical value of the expression.

$$\frac{x^3 \cdot y^{-3} \cdot m^6 \cdot n^3}{x^4 \cdot m^3 \cdot n} \qquad \text{simplifies to} \qquad \frac{1}{\left(x \cdot y^3\right)} \cdot m^3 \cdot n^2$$

$$x := 2 \qquad y := 1 \qquad m := 3 \qquad n := -2 \qquad \text{variable definitions}$$

$$\frac{x^3 \cdot y^{-3} \cdot m^6 \cdot n^3}{x^4 \cdot m^3 \cdot n} = 54 \qquad \text{original expression and value}$$

$$\frac{1}{\left(x \cdot y^3\right)} \cdot m^3 \cdot n^2 = 54 \qquad \text{simplified result and value}$$

Figure 7

For more complicated expressions, the use of simplification and expansion can be used to express the results in clearer form. Figure 8 shows the application of the symbolic processor to the expansion of a polynomial divided by a monomial (case A) and a polynomial divided by a polynomial (case B). Case B is not that successful as the final expression is not simplified from the original. Rather than expand the expression we might try to simplify it. Case C shows the expression from case B with the Simplify option applied to it.

case A	case B	case C
$\dfrac{6 \cdot x^2 + 3 \cdot x + 1}{3 \cdot x}$	$\dfrac{x^2 + 3.5 \cdot x - 2}{x + 4}$	$\dfrac{x^2 + 3.5 x - 2}{x + 4}$
expands to	expands to	simplifies to
$2 \cdot x + 1 + \dfrac{1}{(3 \cdot x)}$	$\dfrac{1}{(x+4)} \cdot x^2 + \dfrac{3.5}{(x+4)} \cdot x - \dfrac{2}{(x+4)}$	$x - .5$

Figure 8

We will further apply these algebraic tools to the factoring of expressions and to the solution of equations and transposition of formulas in Section 2.2.

Explorations

The Basics

In all the following exercises, choose a suitable expression format for the result.

1. Simplify the following expressions:
   a) $6x^2 + 2x + a - 7x + 3a + x^2$
   b) $(a^2 - 4ay) + (6ay - 2a^2) + (7a^2)$
   c) $3x^2 + \{7a - [-2x^2 - (6a - x^2) + (3x^2 - 7a)] + x^2\}$
   d) $(3.5a^2 - 2.7b^2) + (6.9b^2 - 3.2a^2) - (b^2)$
2. Expand the following expressions:
   a) $(ax + b)(cx + d)$ b) $(ax + b)^3$
   c) $(ax + b)^5$

d) $(6.92b^2 + 3.73a^2)(2.75b^2 - 3.50a^2)^2(2.50a^2 + 5.34b^2)$

e) $(x^7 + 3a^2x^4 + x)(x^5 + 2x^3 - x - 2)$ f) $([[[(x)^2 + y]^2 + z)^2 + m]^2 + n)^2$

3. Substitute the expression $(x^2 + 2ax + 2)$ into the z variable of the expression $(z + b)^2 + 2z$ and expand.

4. Substitute the expression $(x + a)^3$ into the b variable of the expression $(z + b)^2 + 2z$ and expand.

5. Simplify the following expressions:

a) $\dfrac{a^3(a+b)^4}{(a+b)^{-2}} \cdot \dfrac{a^{-4}(a+b)^6}{a^7 \cdot a^{-2}}$

b) $(x^2 + y)^3 \cdot z^4 \cdot (m+n)^{-5} \cdot \dfrac{(z^3)^4}{(m+n)^4}$

c) $\left[\dfrac{2x^3 \cdot (3ax)^4 \cdot (-2m^2)^{-3}}{10(ax)^3 \cdot m}\right]^7$

6. Expand and/or simplify the following expressions:

a) $\dfrac{16x^4 - 6x^3 + 2x^2}{2x}$

b) $\dfrac{8x^3a^2 - 3xa^2 + 2x^2a}{-xa}$

c) $\dfrac{46.5x^4 + 0.41x^2a^2 - 8.64a^2}{6.2x^2 + 2.7a^2}$

Beyond the Basics

1. Examine the expansion of the expression $(a + b)^n$ as n takes on positive integer values from $n = 1$ to $n = 7$. What do you notice about the number of terms in the expansion? Is there a pattern to the powers of a and b within each term? Can you find a pattern to the generation of the coefficients of each term? If you can, predict the coefficients for the expansion of the expression with $n = 10$.

2. Verify the following laws of exponents:

a) $(x+a)^n \cdot (x+a)^m = (x+a)^{n+m}$

b) $\dfrac{(x+a)^n}{(x+a)^m} = (x+a)^{n-m}$ for $(x+a) \neq 0$

c) $\dfrac{(x+a)^n}{(x+a)^n} = 1$ for $(x+a) \neq 0$

d) $[(x+a)^n]^m = (x+a)^{n \cdot m}$

and give numerical examples for each case.

2.2 Factoring Using Symbolic Algebra

The simplification and expansion processes you have examined within Section 2.1 contained intermediate steps which were transparent within the process. That transparency is not a bad feature. In some cases, the tools you use become transparent as you gain familiarity and fluency with them. In other cases, the tools need to become transparent as they are only part of a much larger solution strategy.

How is the process of factoring useful? In an age of instant answers (at least, mathematical ones), is there any difference whether an expression is factored or simplified anymore? Does a messy, unforgiving expression take that much more time to calculate than its simplified version? Good discussion topic! Let's examine one case where the process of factoring creates new insight.

In the theory of simple DC (Direct Current) networks, resistors placed in series with a DC source (typically a battery or DC power supply) have a total voltage drop across them equal to the sum of their individual voltage drops. As the voltage across each resistive element is given by Ohm's Law to be $V = I \cdot R$ where I is the current (electron flow) through the element of resistance R then $V_{total} = V_1 + V_2 + V_3$.

However, in a series circuit the current through each element is the same. The flow out of the first resistor must flow into the second and so on. Therefore, since $V_1 = I \cdot R_1$, $V_2 = I \cdot R_2$ and $V_3 = I \cdot R_3$ then the right-hand side of the equation for V_{total} can be expressed as:

$$V_{total} = I \cdot R_1 + I \cdot R_2 + I \cdot R_3 .$$

Since I is a common factor, then $V_{total} = I \cdot (R_1 + R_2 + R_3)$. The sum of the individual resistances can be defined as an effective resistance as measured by the $V_{total} = I \cdot R_{effective}$. Factoring has produced physical insight.

Warmup

The process of factoring usually starts with simple factors, the extraction of a common element from each term of an expression. For the expression, $3x + 9$, the factor 3 is divided out of each term. The factored expression is then $3(x + 3)$. The validity of the factors can always be checked by reversing the process and expanding.

If you attempt to factor the above expression (and you should) then Mathcad will probably return an error message. To factor $3x + 9$, type the expression in an open worksheet making sure to include all arithmetic operators.

☑ In Mathcad 5.0, load the Symbolic Processor before continuing. In Mathcad 6.0, the Symbolic Processor is loaded automatically upon startup.

Select the whole expression by clicking within the mathematical region. A vertical bar will appear. Then press the UP arrow repeatedly or the Spacebar once. Select Symbolic/Factor Expression from the menu. A variety of like-examples is given in Figure 1. Notice that the output has not changed from the input. You might then try simply selecting the variable x. You come up with the same null result. Not to worry, factoring does exist. The program is simply having trouble determining what there is to work with. The common factor here is a number and not a symbol. Luckily, expressions of this level of difficulty are probably faster accomplished using pencil and paper.

(a)	(b)	(c)
$2 \cdot x + 2$	$20 \cdot x + 10$	$30 + 60 \cdot y + 30 \cdot x$
by factoring, yields	by factoring, yields	by factoring, yields
$2 \cdot x + 2$	$20 \cdot x + (2) \cdot (5)$	$30 + 60 \cdot y + 30 \cdot x$

Figure 1

Notice that one of the numbers has been decomposed into factors. The Symbolic factoring process can be applied to whole numbers. Simply type out the number, select it then apply Symbolic/Factor Expression to it. Figure 2 shows some examples of whole numbers being factored.

(a)	(b)	(c)
386400	1056	123009456
by factoring, yields	by factoring, yields	by factoring, yields
$(2)^5 \cdot (3) \cdot (5)^2 \cdot (7) \cdot (23)$	$(2)^5 \cdot (3) \cdot (11)$	$(2)^4 \cdot (3) \cdot (661) \cdot (3877)$

Figure 2

Let's take a look at a more complicated example, one with a common symbolic factor clearly defined. Figure 3 shows the expression. The variable *x* was selected and, from the Symbolic menu, the Collect on Subexpression tool was used. This utility extracts (factors) all instances of the selected variable or expression and its powers from the larger expression.

(a)	(b)
$3 \cdot x + 4 \cdot a \cdot x - 5 \cdot x^2 - b \cdot x^2$	$x^2 + 5 \cdot x \cdot y + y \cdot x^2 + 4 \cdot z \cdot y - 2 \cdot b \cdot y$
by collecting terms, yields	by collecting terms in y, yields
$(-5 - b) \cdot x^2 + (3 + 4 \cdot a) \cdot x$	$\left(5 \cdot x + x^2 + 4 \cdot z - 2 \cdot b\right) \cdot y + x^2$

Figure 3

☑ If an expression cannot be simplified or factored, Mathcad will simply return the expression unchanged. This indicates that factors containing whole numbers were not available. However, if the intended simplification method does not match the type of variable selection you have defined, you will receive an error message.

The next level of factoring applies decomposition rules to quadratics, trinomial expressions of the form $ax^2 + bx + c$ generated from the product of two binomial expressions.

In the case where $a = 1$, if the quadratic expression can be factored to $(x + d)(x + e)$ then the expansion of these factors generates in turn $x^2 + (d + e)x + de$. The conditions for the factoring of the quadratic expression must be $b = d + e$ and $c = d \cdot e$.

Figure 4 shows a variety of quadratic expressions with differing signs for the coefficients b and c. Notice that if whole numbers do not exist to satisfy these two decomposition conditions, Mathcad tries to find rational solutions. If these do not exist, the quadratic is returned unchanged.

(a)	(b)	(c)	(d)
$x^2 + 15 \cdot x + 56$	$x^2 + 8 \cdot x - 20$	$x^2 - 4 \cdot x - 12$	$x^2 - 5 \cdot x + 6$
by factoring, yields	by factoring, yields	by factoring, yields	by factoring yields
$(x + 8) \cdot (x + 7)$	$(x + 10) \cdot (x - 2)$	$(x + 2) \cdot (x - 6)$	$(x - 2) \cdot (x - 3)$

(e)	(f)	(g)	(h)
$x^2 + 2 \cdot x + 3$	$x^2 - 6 \cdot x + 9$	$x^2 + 138 \cdot x + 4472$	$x^2 - 65.2 \cdot x + 1056$
by factoring, yields	by factoring, yields	by factoring, yields	by factoring, yields
$x^2 + 2 \cdot x + 3$	$(x - 3)^2$	$(x + 86) \cdot (x + 52)$	$\dfrac{1}{5} \cdot (x - 30) \cdot (5 \cdot x - 176)$

Figure 4

If you examine the quadratic expressions above in terms of the sign combinations of b and c, a pattern should emerge. This property will be examined within the Explorations.

A general quadratic has a slightly more complicated decomposition due to the nature of its coefficients and constant. If the quadratic expression factors into two binomial expressions *(ax + b)* and *(cx + d)* then these factors must produce $(ax + b)(cx + d) = acx^2 + (bc + ad)x + bd$.

The process of decomposing the coefficient of x, on paper, is the search for the two numbers which add to $(bc + ad)$ yet multiply to $(ab)(cd)$. Figure 5 illustrates this process.

quadratic expression ...	$acx^2 + (bc + ad)x + bd$
decomposition of middle term ...	$acx^2 + bcx + adx + bd$
grouping of terms...	$(acx^2 + bcx) + (adx + bd)$
factoring of groups...	$cx(ax + b) + d(ax + b)$.
yields ...	$(ax + b)(cx + d)$

Figure 5

Figure 6 includes some examples of general quadratics and the application of the Symbolic/Factor Expression tool. You can expand each of the factored expressions to check its validity.

(a)

$-12 x^2 + 7 \cdot x + 12$

by factoring, yields

$-(4 \cdot x + 3) \cdot (3 \cdot x - 4)$

(b)

$26 x^2 + 51 \cdot x - 35$

by factoring, yields

$(2 \cdot x + 5) \cdot (13 \cdot x - 7)$

(c)

$7.5 x^2 - 14.6 x + 6.4$

by factoring, yields

$\dfrac{1}{10} \cdot (3 \cdot x - 2) \cdot (25 \cdot x - 32)$

(d)

$36 x^2 - 36 x + 9$

by factoring, yields

$9 \cdot (2 \cdot x - 1)^2$

Figure 6

The factoring process can be applied to higher order expressions in the same manner. Figure 7 shows the factoring of cubic and quartic expressions.

(a)

$x^3 + 8$

by factoring, yields

$(x + 2) \cdot \left(x^2 - 2 \cdot x + 4 \right)$

(b)

$24 x^3 - 19 x^2 - 62 \cdot x + 8$

by factoring, yields

$(x - 2) \cdot (3 \cdot x + 4) \cdot (8 \cdot x - 1)$

(c)

$-16.5 x^4 - 77.5 x^3 - 114.0 x^2 - 54.0 x - 8$

by factoring, yields

$\dfrac{-1}{2} \cdot (11 \cdot x + 4) \cdot (3 \cdot x + 1) \cdot (x + 2)^2$

(d)

$x^3 - a^3$

by factoring, yields

$-(a - x) \cdot \left(a^2 + x \cdot a + x^2 \right)$

Figure 7

In the process of simplifying algebraic fractions in Section 2.1, common elements or expressions within the numerator and denominator were determined and used to simplify the overall expression. As an example, the fraction $\dfrac{x^2 + 15 x + 56}{x + 7}$ will simplify to $(x + 8)$ since the numerator is the expansion of $(x + 8)(x + 7)$.

Figure 8 shows the steps in the simplification side-by-side with the more detailed steps of first factoring then simplifying.

(a)	(b)	(c)
select the numerator only	select the whole expression	use Simplify
$$\frac{6 \cdot x^2 - 20 \cdot x + 16}{3 \cdot x - 4}$$	$$\frac{6 \cdot x^2 - 20 \cdot x + 16}{3 \cdot x - 4}$$	$$\frac{6 \cdot x^2 - 20 \cdot x + 16}{3 \cdot x - 4}$$
by factoring, yields	by factoring, yields	simplifies to
$$\frac{(2 \cdot (3 \cdot x - 4) \cdot (x - 2))}{3 \cdot x - 4}$$	$2 \cdot x - 4$	$2 \cdot x - 4$
which simplifies to		
$2 \cdot x - 4$		

Figure 8

In case (a), only the numerator was selected to be factored. An intermediate output was generated which clearly indicates the expression common to the numerator and denominator. Then Symbolic/Simplify Expression was used to generate the final output.

In case (b), the same result was obtained by applying Symbolic/Factor Expression while in case (c) Symbolic/Simplify was applied to the whole expression.

The last area differs from the previous ones by focusing attention on the denominator exclusively. This process of Partial Fractions is the reverse of the familiar process of addition of rational fractions. Instead of finding the Lowest Common Denominator and combining two or more fractions into a single expression, this decomposition technique answers the question "What fractions were added to create this simplified fraction?"

The technique is used in Calculus as a tool by which a complicated integration (a type of summation) can be broken down into two or more simpler integrations. And, in the study of time and rate dependent equations, this tool is used in the algebraic manipulation of solutions to these equations (the Laplace Transform).

Figure 9 shows two examples of Symbolic/Convert to Partial Fraction applied to the denominator of each expression. Select only the variable within the expression's denominator.

(a)	(b)
$$\frac{16 - x}{x^2 - 1}$$	$$\frac{4 - 7 \cdot a}{2 \cdot a^2 + 5 \cdot a + 2}$$
expands in partial fractions to	expands in partial fractions to
$$\frac{15}{(2 \cdot (x - 1))} - \frac{17}{(2 \cdot (x + 1))}$$	$$\frac{5}{(2 \cdot a + 1)} - \frac{6}{(a + 2)}$$

Figure 9

Explorations

The Basics

1. Use Symbolic/Factor Expression or Symbolic/Collect on Subexpression (for x) to factor the following expressions:
   a) $35x^2 + b^2ax + bx + 3x$
   b) $-6.0x^2 + 2.5x + 14.0$
   c) $a^2x + bx^3 + cx + ax^3 + x$
   d) $-72x^3 + 42x^2 + 60x$
   e) $16x^4 - 160x^3 + 600x^2 - 1000x + 625$
   f) $27x^5 + 9x^3 + 54x^4 - 46x^2 - 36x - 8$

   In some cases the effect on the expression of either of these tools produces the same result.

2. In Figure 4, various combinations of signs were used for the coefficients of x and the constant. Analyze the examples and generate a rule-of-thumb that would allow you to predict the signs for any quadratic expression of the form.

3. Determine the factors of the following expressions:
   a) $a^2x^2 + 2abx + b^2$, and
   b) $a^2x^2 - 2abx + b^2$ where a and b are constants.

   How would you identify each of these forms as perfect squares before factoring?

4. Determine the factors for
   a) the sum of two cubes, $x^3 + a^3$
   b) the difference of two cubes, $x^3 - a^3$

5. Use the Symbolic/Convert to Partial Fraction utility to determine the fractions which were added together to make up
   a) $\dfrac{-14 - x}{x^2 - 2x - 8}$
   b) $\dfrac{3x + 4}{x^2 + 3x + 2}$

Beyond the Basics

1. In Section 2.1, the Symbolic/Simplify command allowed a simpler expression of algebraic fractions. How would you determine beforehand whether or not the Simplify command (applied once) would generate a final expression different from the original? (The Symbolic/Factor Expression tool can be applied in turn to each of the numerator and denominator within the algebraic fraction.) Apply your reasoning to:
   a) $\dfrac{2x^5 + 6x^4 + 2x^3 - 8x^2 - 4x + 2}{x^2 - 1}$

   b) $\dfrac{x^3 + 12x^2 + 47x + 60}{x^2 + 3x + 2}$

2. In DC circuit theory, the voltage applied to a network of parallel resistors acts equally on each of the resistors. The total current is split between the pathways offered by the resistors. Each branch's current follows Ohm's Law. In the first resistor then, $I_1 = V/R_1$. And so on for all other resistors parallel to this one. If the total current is given by $I_{total} = I_1 + I_2 + I_3$, show that it is possible to define an effective resistance $R_{effective}$ such that $\dfrac{1}{R_{effective}} = \dfrac{1}{R_1} + \dfrac{1}{R_2} + \dfrac{1}{R_3}$.

3. Use Symbolic/Convert to Partial Fraction to decompose the following algebraic fractions:
   a) $\dfrac{34ax + 153a - 24bx + 16b}{6x^2 + 23x - 18}$
   b) $\dfrac{29x^2 + 6x + 13}{x^3 - 3x + 2}$

2.3 Solving Equations

In this set of explorations, we will examine the application of symbolic solution tools to the solution of literal and numerical equations and expressions. The underlying routines which generate symbolic answers are variations of the pencil-and-paper techniques you have learned to isolate variables within an expression or to solve a literal or numerical equation.

Warmup

An algebraic relation or equation describes a particular relation between the elements to the right and left of the conditional operator ($=$, $<$, $>$,...). There is a resemblance to the structure of a typical English sentence with its subject, verb and predicate. Just as it is possible to express the idea of the sentence in many different ways by reordering its elements, a mathematical statement can be reordered without affecting the truthfulness of the condition as long as the reordering process keeps the condition true at each step. This is the basis of any transposition techniques (addition, division, factoring,...) you have learned using pencil and paper.

In the equation $a = b \cdot c$, the variable b can be isolated by applying the same operation to both sides of the equation simultaneously, namely by dividing both sides by the variable c provided $c \neq 0$. Thus, $b = \dfrac{a}{c}$. This process of balanced operation can be applied to more complicated expressions with the same efficiency.

Here we are looking at the use of the algebraic solution tool as an intermediary step in a longer process and so will not dwell on how the solution was derived . The 'how' of the derivation is covered in more detail in your pencil and paper work.

For the expression $a = b \cdot c,$ open a worksheet in Mathcad and move the cursor to a central location.

    ☑ In Mathcad V5.0 you will have to load the symbolic processor before proceeding. Select Symbolic/Load Symbolic Processor from the menu.

In numerical calculations, the equal sign (=) is a request for Mathcad to perform the numerical calculation with the predefined variables and to output a result. In Automatic mode the result is output as soon as you finish pressing the equal sign. If Automatic mode is not selected, then press F9. In symbolic manipulations, another type of equal sign must be defined. The bold-faced equal sign is generated by pressing CTRL-=.

For our expression, type a {CTRL-=} $b \cdot c$ (note: without the brackets). Then select b by first clicking beside the variable then pressing the UP arrow once.

If you have trouble selecting the variable $b,$ try clicking on the other side of the variable and pressing the UP arrow. Don't get frustrated. Even veteran Mathcad users see the selection process as an art form!

From the Symbolic menu, select Symbolic/Solve for Variable. Figure 1 shows the result of this process with the Derivation Format set at 'Show derivation comments' selected and the 'horizontally' radio button turned on.

the symbolic equal sign is CTRL = and appears in **bold**

$$a \boldsymbol{=} b \cdot c \qquad \text{has solution(s)} \qquad \frac{a}{c} \qquad \text{... for the variable b}$$

Figure 1

One of the difficulties of using this tool is the lack of any reference within the comments as to exactly which variable you have solved for or have isolated. This confusion can be avoided by getting into the habit of including a short explanatory note alongside the output.

Figure 2 shows the isolation of each variable in turn within a more complicated expression.

$$(a)\ 3 \cdot x \boldsymbol{=} \frac{4 \cdot y + z}{5 \cdot b} \quad \text{has a solution for y of} \qquad \frac{5}{4} \cdot \left[3 \cdot x - \frac{1}{(5 \cdot b)} \cdot z \right] \cdot b$$

$$(b)\ 3 \cdot x \boldsymbol{=} \frac{4 \cdot y + z}{5 \cdot b} \quad \text{has a solution for b of} \qquad \frac{-1}{15} \cdot \frac{(-4 \cdot y - z)}{x}$$

$$(c)\ 3 \cdot x \boldsymbol{=} \frac{4 \cdot y + z}{5 \cdot b} \quad \text{has a solution for z of} \qquad 5 \cdot \left[3 \cdot x - \frac{4}{(5 \cdot b)} \cdot y \right] \cdot b$$

$$(d)\ 3 \cdot x \boldsymbol{=} \frac{4 \cdot y + z}{5 \cdot b} \quad \text{has a solution for x of} \qquad \frac{4}{(15 \cdot b)} \cdot y + \frac{1}{(15 \cdot b)} \cdot z$$

Figure 2

And, as Figure 3 demonstrates, the isolation of variables may yield more than one answer depending on the degree of the variable.

$$a)\qquad a \boldsymbol{=} b^2 + 2 \cdot a \cdot c + d \qquad \text{has solution(s)} \qquad \frac{-\left(-b^2 - d\right)}{(1 - 2 \cdot c)} \qquad \text{...for a}$$

$$b)\qquad a \boldsymbol{=} b^2 + 2 \cdot a \cdot c + d \qquad \text{has solution(s)} \qquad \begin{pmatrix} -\sqrt{a - 2 \cdot a \cdot c - d} \\ \sqrt{a - 2 \cdot a \cdot c - d} \end{pmatrix} \qquad \text{...for b}$$

Figure 3

The results of Figure 3 can be evaluated numerically by first defining the variables. From the expressions, variables *a, c* and *d* were defined as $a = 6$, $c = 2$ and $d = -20$. The whole of the output was selected by clicking on the area and pressing the UP arrow (or Spacebar). A new variable *B* was defined in terms of the selected region by copying and pasting the selected region to the right-hand side of the definition expression. Then *B* was evaluated by typing $B = $.

for the equation $\qquad a \boldsymbol{=} b^2 + 2 \cdot a \cdot c + d \qquad$ define $\ a := 6 \quad c := 2 \quad d := -20$

paste the expression for b into B := and then evaluate as a floating point number

$$B := \begin{pmatrix} -\sqrt{a - 2 \cdot a \cdot c - d} \\ \sqrt{a - 2 \cdot a \cdot c - d} \end{pmatrix} \qquad\qquad B = \begin{pmatrix} -1.4142 \\ 1.4142 \end{pmatrix}$$

Figure 4

You may be tempted to use the Symbolic/Evaluate Expression to perform the evaluation once the variables have been defined. In this way, you would not have to define the intermediate *B*. However, this works only for outputs which have first been generated from numerical equations. Figure 5 illustrates the difference. The first output was generated from the equation with variables a, *c* and *d* defined within. The variable *b* was selected and solved for. This produced an exact answer which was then converted to a floating point output by selecting Symbolic/Evaluate/Floating Point Evaluation... for a floating point precision of 5 decimal places.

In the literal expression $a = b^2 + 2 \cdot a \cdot c + d$, values have been assigned
to a, c and d. An exact or a floating point result can be determined

$6 = b^2 + 2 \cdot 6 \cdot 2 + (-20)$ has solution(s) $\begin{pmatrix} -\sqrt{2} \\ \sqrt{2} \end{pmatrix}$

floating point evaluation yields $\begin{pmatrix} -1.4142 \\ 1.4142 \end{pmatrix}$

Figure 5

☑ The distinction between isolating a variable and solving an equation may not be a clear one. It may not even be that useful and may depend on your future intentions. In either case, the operation used within Mathcad is the same. The only distinction may be in the structure of the original equation or expression, whether it be a pure literal or a numerical expression.

The solution of the quadratic equation is an especially nice result of the application of the completion of squares technique. The pencil and paper solution not only gives you an appreciation for the history of mathematics but also for the use of the word 'square'.

The general quadratic formula is $ax^2 + bx + c = 0$ where $a \neq 0$. If you select the variable *x* and select Symbolic/Solve for Variable, the output generated should look like that in Figure 5. You can confirm that this result is exactly the one you expected from the quadratic formula.

---

literal quadratic equation

$a \cdot x^2 + b \cdot x + c = 0$ has solution(s) $\begin{bmatrix} \dfrac{1}{(2 \cdot a)} \cdot \left(-b + \sqrt{b^2 - 4 \cdot a \cdot c} \right) \\ \dfrac{1}{(2 \cdot a)} \cdot \left(-b - \sqrt{b^2 - 4 \cdot a \cdot c} \right) \end{bmatrix}$

these results correspond to those of the quadratic formula

Figure 6

If you have a numerical quadratic equation (with the coefficients and constant defined) the exact numerical solution can be generated by including values for *a, b* and *c* within the equation and then proceeding as before. If the exact numbers output are to be transformed into floating point estimates, then selecting Symbolic/Evaluate/Floating Point (with the floating point precision set at 5) produces the output in Figure 7.

numerical quadratic equation with values assigned to a, b, c

$$3 \cdot x^2 + 4 \cdot x - 10 = 0 \qquad \text{has solution(s)} \qquad \begin{bmatrix} \dfrac{-2}{3} + \dfrac{1}{3} \cdot \sqrt{34} \\[2mm] \dfrac{-2}{3} - \dfrac{1}{3} \cdot \sqrt{34} \end{bmatrix}$$

floating point evaluation yields $\qquad \begin{pmatrix} 1.2769 \\ -2.6103 \end{pmatrix}$

Figure7

Explorations

The Basics

1. Re-express the equation $I = P \cdot r \cdot t$ for each of *P, r* and *t.*

2. Solve the linear equation $\dfrac{4x}{5} + 2x = \dfrac{9}{15}$.

3. Solve the equation $y^2 + 2y = 3$.

4. Solve the quadratic equation $3x^2 + 4x - 10 = 0$.

5. Show that the equations $3x^2 + 2x = 4 + 3x$ and $3x^2 - x - 4 = 0$ each have the same solutions.

6. Solve the equation $x - 7 = 3x - 6x + 8$. Determine the floating point estimate of the answer. If the expression $2z$ were added to the right-hand side of the equation, how would the answer change?

7. In a simple DC circuit, the voltage (*V*) across two resistors *R* and *r* connected in parallel is given by $V = I \cdot \dfrac{Rr}{R+r}$ where *I* is the total current across both resistors.

   a) Determine the expression for *r* in terms of the other variables.

   b) Evaluate *I* for $V = 25$ Volts, $R = 100$ Ohms and $r = 200$ Ohms.

8. The force of gravity between two massive objects is defined by $F = \dfrac{GMm}{r^2}$ where *G* is the universal gravitational constant, *M* and *m* define the masses and *r* the distance between the masses. Express *r* in terms of *F, G, M* and *m.* Are both solutions for *r* allowed? Explain.

9. When invested money is allowed to earn compound interest, the total amount accumulated after a number of compounding periods is given as $S = P(1 + r)^n$ where *S* is the sum, *P* is the original amount invested (the principal) , *r* is the periodic rate of interest (yearly interest divided by number of compounding periods in a year) and *n* is the number of compounding periods.

   a) Determine the expression for n in terms of the other variables.

   b) How many years would it take for a principal of $10,000.00 to grow to $20,975.68 if the interest rate were 10% per year compounded quarterly (i.e., every 3 months).

10. The total resistance (*R* in Ohms) for two resistors in series is $R = \dfrac{\rho L_1}{A_1} + \dfrac{\rho L_2}{A_2}$ where *L* is the length (m) of each resistive element, *A* is the cross-sectional area (m^2) and ρ is the resistivity (Ohms·m). Determine the expression for ρ in terms of the other variables.

Beyond the Basics

1. Two equations are given by $y = 3x^2 + 4x$ and $y = 5x + 10$. Determine the values of x common to each equation which generate common values of y. Visually, these common values of x and y represent the points at which the two plots intersect on the plane.

2. As a long solid rod expands or contracts due to changes in temperature, its length can be predicted using $L = L_0 (1 + \alpha \, \Delta T)$ where L_0 is the original length, ΔT is the change in temperature and α is the coefficient of linear expansion. Determine an expression for α. Determine the coefficient of linear expansion if a temperature change of 40°C causes a 1% change in the length of a 1.5 m rod.

3. The area expansion formula is derived from the square of the length expansion formula.

    For $A = [L_0 (1 + \alpha \Delta T)]^2$,

    then $A = A_0(1 + 2\alpha\Delta T + \alpha^2\Delta T^2)$.

    where A_0 is the original area defined by L_0^2.

    Determine the new expression(s) for α in terms of the other variables of the area expansion formula.

4. Solve the equation $xy^3 - 1 = 0$ by selecting y and using Symbolic/Solve for Variable.

5. For one-dimensional dynamics of a mass in a gravitational field, the equation of distance and time is given by $d = d_0 + v_0 t - \dfrac{1}{2} g t^2$. Here, d_0 is the original height of the mass above the ground, v_0 is the initial velocity in the vertical direction and g is the acceleration due to gravity. Determine the relation for the time of flight of the mass (i.e. the difference between the times at which $d = 0$).

6. The cubic equation, $x^3 + 5x^2 - 2x + 3 = 5$, has a symbolic solution which scrolls off the page due to its length and complexity. Comment on the strengths and weaknesses of symbolic solutions applied to problems like this.

7. Attempt to solve the equation $x = \cos(x)$ using the symbolic tools available. How else could you approach the solution to this problem?

Chapter 3: Functions and Graphs

The idea of a function, of a one-to-one correspondence between the input of an independent variable and the subsequent output of the dependent variable, is at the heart of all mathematically described systems.

Although the advent of hand-held calculators made the tedious task of determining a function's coordinates less painful, only lately has the ready and cheap supply of programmable graphics calculators and plotting programs made the visualisation of these functions accessible to all.

This unit will expose you to a variety of function types and representation styles. The syntax of the functional expression and its display in tables and graphs will be examined.

In the new Mathematical reform, attention is being focussed on the expression of mathematics in all its forms: symbolic, numerical and graphical. As a mathematical explorer, you will find that Mathcad offers all of these avenues by skillfully blending robust numerical techniques, highly editable graphical regions and a subset of Maple's Symbolic Processor program.

3.1 Functions, Graphs and Tables

A device which merely absorbed input would be the ultimate waste disposal unit. Unfortunately, the design would probably violate a few natural conservation laws.

In this exercise, we will examine connections, inputs creating outputs in a definable and usable fashion. The creation and understanding of these connections allows us to visualise them, extrapolate and interpolate them. Most importantly, the process allows us to deduce patterns useful to projection and simulation.

From an operational standpoint, the process of a function takes an input value, transforms it through the recipe defined by the functional relationship, and outputs a value which is specific to that particular input. There is a one-to-one correspondence of the independent and the dependent variables. The device outputs one particular output for one particular input.

There are two different directions you can take here.

The first is to represent an already defined function's behavior with tables of data points (the inputs and outputs) or with plots. These could be compared to experimental data. In the case of the linearity of Ohm's Law, the expected plot or table could be determined based on the value of the resistor and the laboratory data compared to the 'expected' results.

The second, and more involved path, is to start with raw data, data points which you feel should be connected in some fashion and impose a functional relationship on them. This will be the focus of a further set of explorations.

In this set of exercises, we will examine Mathcad's structure for handling functions, the definitions of inputs and outputs, and the many user-defined formats these can be represented in.

Warmup

So far you have used Mathcad to perform rudimentary calculations and to define expressions, constants and variables. The inputs have been fixed as have the results. Now, we will allow more flexibility in the inputs, allow them to cover a range of values and let the output track this succession.

Open a new Mathcad document and define a range of input values for x over [-3, +3]. The range symbol (..) is a result of pressing the semi-colon. Define the function $f(x)=x^2$ - 2. The definition symbol ': = ' is created by typing the colon. Include any necessary documentation.

$$x := -3 .. 3$$
$$f(x) := x^2 - 2$$

Figure 1

☑ Mathcad will not recognize the definition symbol typed as ':' followed by '='.
   The range specifier cannot be typed as '. .' (i.e., period, period).

To create a rectangular plot region, scroll down the screen using the cursor keys and type "@" or select Graphics/Create X-Y Plot or click on the rectangular plot icon from the Graphing Palette. A plot region will appear on your screen with empty axes label and limit holders defined by a ■.

The "missing operand" error signal in Figure 2 suggests that you have to input a function or other data-set name into the offset vertical axis placeholder before the region becomes active. Along the horizontal axis, the middle place holder represents the name of the independent variable or data set. The remaining placeholders contain the plot limits which are either program- or user-defined.

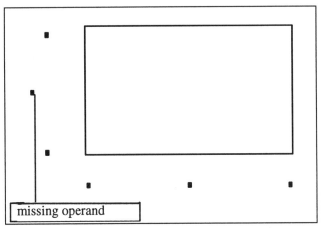

Figure 2

Label the x and y axes using either the TAB key or the mouse to move from place holder to place holder. Mathcad will substitute its own values if the axes limits are left unfilled. As you move you should see the plot of the quadratic fill the plot region and the axes limits fill. You will probably note that your plot is a quadratic out of a connect-the-dots game.

☑ Mathcad refreshes the screen automatically unless you switch this option off. For long editing sessions, the screen refreshes can be exceedingly annoying and time consuming. In the Math menu, toggle off the Automatic Mode option. From then on, in the Manual mode, press F9 to start the calculation or select Math/Calculate.

How many points were included in your plot? You can examine this by creating data tables for the input and output variables. In any region underneath or to the right of the definitions of x and $f(x)$, type "$x =$" and beside it "$f(x) =$". If nothing appears immediately, press F9.

To increase the number of points used in the table or graph, you need to redefine the step-size of x. By default, this is set at 'one'. Rather than specify the increment size, Mathcad accepts the next number specified in the variable set. To increment your domain by tenths (0.1), edit the definition of x to

$$x := -3.0, -2.9 .. 3.0$$

This tells Mathcad to increment the left limit by 0.1 this time rather than by the default value of 1.0. Recalculate and observe any changes.

☑ For small increments, tables become overwhelming and are self-defeating. If you require a high resolution plot, the small step-size would create a table extending over a few screens. No one benefits by having to read headache-inducing tables. If required, individual output values can be generated by typing "f(xvalue) =" for a particular numerical value. Or, the x range for a small number of inputs and outputs can be redefined for a particular high resolution range after the plot region. See Figure 3 below.

As you finished editing your domain step-size, the plot should have changed in its resolution. The line will now be smoother and more like that you expect. The table will have grown but probably still be of an acceptable size.

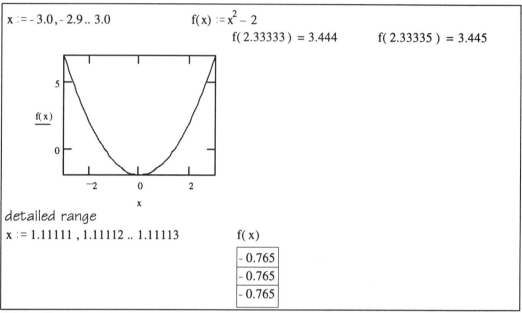

Figure 3

Notice that the formats for the plot and table regions are relatively featureless. These are the default settings. Try decreasing your step-size in *x* to 0.00001 and examine the table outputs. The numerical detail you expected has not been lost. Mathcad uses 15 decimal place accuracy in its calculations. However, that accuracy is being masked by the numerical default format. The plot probably has not changed significantly.

Numerical Formats

Move the cursor into a table region and double-click. A dialog box appears. Select and play with changes to all the settings. You can control the precision (the number of decimal places displayed), the exponential threshold (the scale at which the numbers are represented in scientific notation), and the zero tolerance (the size a number has to be considered a zero). Numbers may also be expressed in decimal, octal, and hexadecimal notation.

A LOCAL format applies to a specific region within your document and overrides any GLOBAL formats.

Edit the dialog box for the LOCAL and GLOBAL formats by selecting the either option. Examine the effects on the display of your numerical design.

The look of your numbers, graphs and text has become a personal choice. No two people working on the same problem will express themselves identically.

Graphical Formats

The size of the graphical region can be adjusted by clicking just outside the region and dragging the dashed selection rectangle over the plot. Within the selection rectangle, the cursor will change to a cross. The plot region as a whole can then be moved by clicking and dragging to a new location. Along the bottom and right-hand borders the cursor will change to a resizing arrow. The size of the plot can be adjusted by clicking and dragging this arrow.

If you wish to change the design of your plot region, double-click on a plot. Another dialog box emerges. Here you have control over the plot type and color (dots, lines, stars, bars, steps), the number of superimposed grid lines, the display of numbers along the axes and the cyclical spacing of the grid lines (rectangular vs. logarithmic or semi-logarithmic).

Examine the effects of each of these format options on the graph of *f(x)*.

Explorations

The Basics

1. With the function in the Warmup (or any other you like), show five separate and distinctly different locally formatted tables of the same data.
2. Again, using the same function or another of your creation, display five separate and noticeably different styles of graphical representation.
3. Produce a file which contains at least four linear functions and examine the use of color and plot-types on these straight lines. The standard form of a linear function is $f(x) = mx + b$ where m is the slope and b the *y*-intercept.
4. Determine the conditions necessary for functions producing parallel straight lines.
5. Determine the conditions necessary for functions producing perpendicular straight lines. Note that you may have to adjust the size and/or scaling of your graph so that a right angle intersection appears correctly.
6. When you select a plot region, an X-Y Plot menu option appears along the menu line at the top of the active window. Create a function which crosses the x-axis in at least one location. Select the Zoom option in the drop-down menu and estimate the location(s) at which the function crosses the *x*-axis.

Beyond the Basics

1. Laboratory tests of passing current through a 100 Ohm resistor create the following current and voltage (I,V) data pairs (in amperes, volts): (0.042, 5.0), (0.113, 10.0), (0.129, 15.0), (0.212, 20.0). The individual data points can be entered directly into the Mathcad worksheet. First create a dummy index which counts the number of points. Type "k : 1 ; 4". To input the data values for current, type "I [k : 0.042 , 0.113 , 0.129 , 0.212". A table of *I* values will be created with four entries. The *V* data table can be created in the same manner. Plot these data points and compare them to the linear function $I = V \cdot G$ where G is the reciprocal resistance (or conductance).
2. Depending on your hardware, a computer can more quickly plot and tabulate a function if the function is in the form of nested linear factors rather than in the straight polynomial form. The definition of the function remains unchanged. However, the computation is handled differently. Then, $f(x) = x^3 + 2x^3 - 6x + 2$ can be rewritten as $f(x) = ((x + 2)x - 6)x) + 2$ without any change in the output values. Examine the effect on computation time for this or another function. You may have to use a very fine step-size, in which case you do not want a table represented.

3.2 The Root Function and Solve Blocks

The solution of quadratic equations has a long history. In the 9th Century A.D., Arab mathematician Al-Khwarizmi produced the first textbook on Algebra, *Al-jabr wa'l muqabalah*, which translates very roughly to 'Restoration and Balancing'.

In the third chapter of his book, Al-Khwarizmi gave the recipes for finding the solutions to three quadratic equations:

1. $x^2 + 10x = 39$
2. $2x^2 + 10x = 48$
3. $\frac{1}{2}x^2 + 5x = 28$

The solution he presented to the problem of the quadratic equation is particularly elegant and appeals to those raised on a straight diet of symbolic manipulation. Rather than rely on the symbols, he presented a variable squared as a SQUARE of area *x* by *x*. Of course, that is exactly what a variable squared should be geometrically. Squares and rectangles were drawn to represent the values given within the problem. The area needed to complete the outline of a perfect square was added to the diagram and its value to each side of the equation. The square root of each side then represented the length of the side of the perfect square.

The use of geometry and picture was presented as solution and understanding. In many ways, the use of computers in the analysis of Mathematics has given us back a bit of the 9th Century.

More recently, the solution of quadratic and higher-order equations plays a great part in the analyzed behavior of dynamic systems whether electrical, mechanical or industrial. Systems controlled by their rates of change give rise to dynamic equations whose solutions often contain quotients of polynomials. The zeros of the numerator and denominator can provide information about the natural and forced system responses.

Warmup

The roots of a function can be thought of as the solution(s) to the equation $f(x) = 0$. Those root values of *x* drive the varying function to its intersection with the *x*-axis. Mathcad offers many tools which can be applied to the root.

The Root Function

Mathcad includes a routine which will numerically solve the real roots of a function. It requires an estimate of the root, a seed point from which to start the process and returns a value to the desired accuracy. The details of the method are left for a further exploration using iterative approaches, the Secant Method and Newton's Method.

Depending on the complexity of the function, an estimate of the root may itself require a little work. A simple linear function's root may be obvious but as the functions move toward quadratics and cubics, the behavior of the function is not always obvious. A plot may help you focus on the approximate point.Or, it can save you the effort if in fact there are no roots to the function.

The Root Function may not always return an answer. The cyclical routine may not converge to an answer within the tolerance settings. By default, Mathcad sets the variable TOL to 0.001. As the routine cycles through its steps, the answer from the current cycle is compared to that of the previous answer. If the difference between these is less than the tolerance, the routine ends.

Obviously, the smaller the tolerance setting, the more accurate and reliable your answer is. However, you may have sacrificed some time for this improved accuracy.

For the function $f(x) = x^2 - 3$, the plot of the function reveals two roots, two values of 'x' which drive the function to zero. The points appear to be near $x = -1.7$ for the left root and $x = 1.7$ for the right root. See Figure 1.

Using these estimates, the Root Function can be invoked to solve the equation. First, define an estimate of the particular root, xguess. The form the function takes as a definition is then

$$\text{your\_root} := \text{root}(f(\text{xguess}), \text{xguess})$$

Or, you can output an answer directly by foregoing the definition step. Note that the inputs into the 'root'-ine are names not numbers.

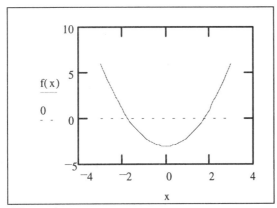

Figure 1

Figure 2 displays the first trial with the Tolerance set to its default value of 0.001. The numerical precision is set at 15 decimal places.

TOL := 0.001 *setting*	xLR1 := -1.7 *estimate*	
leftroot := root(f(xLR1), xLR1)	*root function definition*	
leftroot = -1.732336772790075	*solution*	

Figure 2

In the second trial, Figure 3, the tolerance is decreased or tightened to 10^{-11}. Notice the slight change in the answer. Depending on your hardware, you may not have noticed an appreciable change in the computation time.

TOL := 10^{-11} leftroot2 = -1.732050807568877

Figure 3

For the third trial, Figure 4, the tolerance is decreased to 10^{-15}. There is no appreciable change in the answer.

TOL := 10^{-15} leftroot3 = -1.732050807568877

Figure 4

Try setting the tolerance below this level. How do you balance the need for accurate answers with the need to have an answer sometime in the near future?

Solve Blocks

The Solve Block utility which Mathcad offers is another iterative program driven by the setting of the Tolerance variable. However, this utility can be applied not only to the problem of finding roots but to the solution of any equation or system of equations.

The Solve Block requires a starting point set by the definition of an estimate for the solution. The Tolerance variable needs also to be reset depending on the required accuracy. Figure 5 shows the solution to the same quadratic using this technique. Note that the 'equals' sign in the statement of the equation is typed using CTRL =.

$\text{TOL} := 0.001$ set tolerance $x := -1.7$ estimate

Given   invoke Solve Block routine

$x^2 - 3 = 0$ equation with "Ctrl =" sign

$\text{xroot} := \text{Find}(x)$ define variable to be solved

$\text{xroot} = -1.732050807568877$ left root

Figure 5

Experiment with the Tolerance setting in the above example. This routine seems to be a bit more robust in that the answer is generated to 15 decimal place accuracy with little tolerance effect.

Maple's 'Solve for Variable'

The last routine uses the subset of Maple included in Mathcad. Maple is a powerful symbolic and numerical mathematics program. The two routines we have examined so far work very well for existing real roots. Maple, however, can also provide the imaginary roots of an equation should they exist.

Load the Symbolic Processor from the Symbolic menu. Maple within Mathcad acts on selected variables and equations. After you have typed the expression you wish to find the roots to, select the variable x. Under the Symbolic menu, click on the 'Solve for Variable' option. The roots of the equation will appear under or beside the expression depending on the Derivation Format preferences you have selected within the Symbolic menu.

Figure 6 contains the Maple solutions to the quadratic equation. A symbolic and exact solution has been output rather than a solution requiring a numerical process.

$x^2 - 3$ has solution(s) $\begin{pmatrix} \sqrt{3} \\ -\sqrt{3} \end{pmatrix}$

Figure 6

The next expression, in Figure 7, would return a 'did not find solution' notice if either of the first two methods were used. However, the symbolic engine in Maple encounters no difficulty finding both the real and imaginary roots of the expression.

$$x^3 + 3 \qquad \text{has solution(s)} \qquad \begin{bmatrix} -3^{\left(\frac{1}{3}\right)} \\ \frac{1}{2} \cdot 3^{\left(\frac{1}{3}\right)} - \frac{1}{2} \cdot i \cdot 3^{\left(\frac{5}{6}\right)} \\ \frac{1}{2} \cdot 3^{\left(\frac{1}{3}\right)} + \frac{1}{2} \cdot i \cdot 3^{\left(\frac{5}{6}\right)} \end{bmatrix}$$

Figure 7

Explorations

The Basics

1. Determine the roots of the function $f(x) = x^2 - 5$ using the numerical, graphical and symbolic techniques.
2. Find the roots of the function $f(x) = 2x^3 + 3^2 - 4x - 1$ using each of the three methods discussed in the Warmup.
3. Use graphical and numerical techniques to solve the equation $ax^2 + bx + c = 0$ for each of the following sets of $\{a, b, c\}$:
   a) $a = 4.2, b = 6.5, c = 1.0$
   b) $a = 9, b = 12, c = 4$
   c) $a = 7, b = 2, c = 6$.
4. Projectile height as a function of time is given by $y(t) = y_0 + v_0 \sin(\theta)t - 0.5\, gt^2$. Here, y_0 represents the initial height, $v_0 \sin(\theta)$ is the initial velocity component in the vertical direction and g is the acceleration due to gravity (9.8 m/s^2). Determine the times at which the height of the projectile is zero. What do these times represent? How is the maximum attained height calculated using this information?
5. For a series capacitor-inductor circuit, the reactance generates a quadratic in the angular frequency. The total reactance is given by $X = \omega L - \dfrac{1}{\omega C}$ where X is in Ohms, L is the inductance, C is the capacitance and ω, the angular frequency. For a total reactance of 2500 Ω, determine the angular frequency for $L = 150$ mH and $C = 10\ \mu$F. Edit this file to tabulate the frequency response of two other L-C combinations for the same total reactance.

Beyond the Basics

1. The solutions of *The Basics* #3 can be interpreted as the roots of the function $y = ax^2 + bx + c$ with each of the solutions representing a distinctly different type of intersection of the function with the x-axis. How can the values of a, b and c be used to classify the number of real roots of the function?
2. As mentioned in the Introduction, the solution of polynomial equations is part of the determination of the system response for a dynamic (time-dependent) system subjected to an applied force. The transfer function is a measure of the ratio of the input to the output as measured in the frequency or s domain. For the following transfer or system function, determine the zeros (those values of s which force the numerator to zero) and the poles (those values of s which force the denominator to zero).

$$T(s) = \frac{s^2 + 18s + 20}{5s^3 + 30s^2 + s - 10}$$

3. Create a function which displays more than 5 roots over a limited range. Discuss the merits of and problems in using each of the techniques as your only root solver.

4. The gravitational force of attraction between two objects of masses M and m a distance r apart is $F = \dfrac{GMm}{r^2}$ where G is the Universal Gravitational Constant equal to $6.67\text{x}10^{-11}$ m^3/(kg s^2). A mass of 105 kg is separated by a distance of 5.0×10^8 m from a mass of 103 kg. Determine the distance along the radius joining the two masses at which a third mass of 10 kg would experience no net gravitational attraction.

5. Use the symbolic processor to solve the general quadratic equation, $ax^2 + bx + c = 0$.

6. The solutions to the system of equations $y_1(x) = 5x^2 - 4$ and $y_2(x) = -10x + 2$ are the points each function has in common with the other. As graphs, the solutions are indicated as intersection points. How are the intersection points for $y_1(x)$ and $y_2(x)$ defined by the solution to the quadratic equation, $5x^2 + 10x - 6 = 0$?

References:

1. *A History of Mathematics*, 2nd edition, C.B. Boyer and U.C. Merzbach, John Wiley and Sons, 1991, 227-233
2. *Electric Circuit Theory*, 2nd edition, R. Yorke, Pergamon Press, 45-49

3.3 Exponential Growth and Decay

The general form of an exponential function is $f(x) = b^x$ where b is a non-negative number.

A familiar case of exponential functions applied to the growth of an amount is that of money, at least money which is left untouched over a period of time. If interest is applied to this original amount, the money feeds on itself, growing at a fixed rate.

The same pattern of growth connected to amount may be seen in unchecked bacterial growth, your own learning curve, or the charging and discharging of capacitors.

Unfortunately for money there are many adverse conditions which limit its growth: poor markets, loss of employment, economic depressions, and death. However, adverse conditions should not be seen as always negative. Extreme cold limits the growth of mold in the back of your refrigerator. In natural systems, the predator-prey food supply relation usually sets a limit on the total number of organisms a system can support. In a capacitor, this limit is set by the physical characteristics of the element.

The same mathematical form can be applied to a wealth of situations, to any mechanical, biological, economic, or electrical system where the rate of growth or decay depends on the size of the system.

Warmup

If a percent yearly growth rate (rate) is applied to an initial amount of money (sum0), the new amount or sum as the years progress can be easily calculated. After the first year, the new sum will be given by

$$\text{sum1} = \text{sum0} + \text{sum0·rate}$$

In the second year, the interest rate is feeding not off the initial amount (sum0) but the amount present at the start of its period (sum1). At the end of two years

$$\text{sum2} = \text{sum1} + \text{sum1·rate}$$
$$\text{or}$$
$$\text{sum2} = \text{sum1}(1 + \text{rate})$$

But (sum1 = sum0 + sum0·rate) so

$$\text{sum2} = \text{sum0}(1 + \text{rate})(1 + \text{rate})$$

This process can be continued forever. At the risk of losing patience, a moment's thought shows the pattern underlying the growth. For every year 'sum0' is multiplied by yet another factor (1 + rate). This *compounding factor* defines the growth of the initial amount over any period of time.

After a total period of 't' years, the amount of money present in your account would have grown from the 'sum0' value to

$$\text{sumt} = \text{sum0}(1 + \text{rate})^t$$

The geometric progression of the initial amount then allows a more compact form.

At a rate of 5% per year for 10 years, an initial amount of $1000.00 would grow to $1628.89. Its behavior could be represented by a continuous curve. However, this plot would confuse the fact that the growth is not continuous but rather appears at well defined and discrete intervals. In our case, and as Figure 1 shows, the interest would be paid at the end of each compounding period.

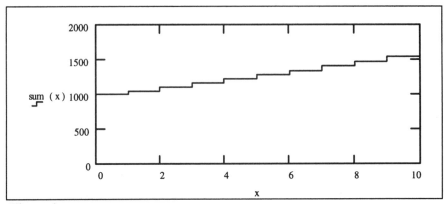

Figure 1

To increase the rate of growth, we could decrease the compounding period from 1 year to 6 months. The interim rate would be half of the original 5% but there would be twice as many periods. Figure 2 shows the growth of the $1000 over the same number of years. The final amount now is $2653.30, over $1000 more than the yearly compounded final total for the same number of years.

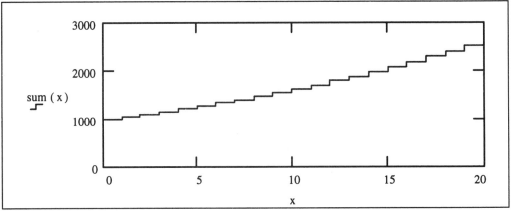

Figure 2

Notice from the above figure that the incremental change in the vertical direction is growing as *x* progresses. More money (or bacteria or people) is (are) making more money (or bacteria or people).

What would the final sum be for a quarterly rate of 0.0125 compounded over the same 10 years?

Clearly, an investment with quick compounding would grow to a substantial amount in no time at all. This is one of the best reasons you will never see this kind of account at a bank.

Now, let's generalize the exponential function to $f(x) = a \cdot b^{kx}$ where *b* is a non-negative number. For $b = 2$ and $a = 1$, allow the *k* value to vary from 1 to 4 for *x* over 0 to 4. Figure 3 shows the function $f(x,k)$ over the domain.

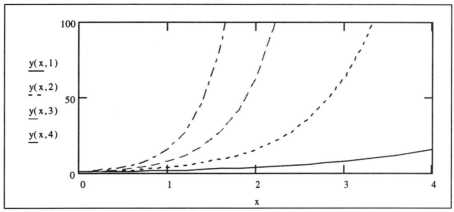

Figure 3

Can compounding become continuous? For quantities such as people, bacteria and electrons any changes to a population happen in discrete jumps. However, relative to a large total population a growth of one discrete unit is difficult to tell apart from a seemingly continuous change. The jumps can then be approximated by a flow. The error involved is minimal and the advantages outweigh the disadvantages.

Let's examine the behavior of the compounding factor $factor(x) = \left(1 + \dfrac{1}{x}\right)^x$ as x grows infinitely large. This expression is consistent with our previous factors since as the number of compounding periods increased, the periodic rate decreased. Figure 4 shows the graphical results of the factor for x from 1 to 100.

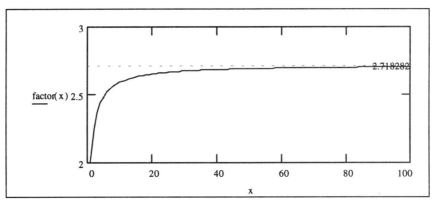

Figure 4

The progression tends toward a limit of roughly 2.718. This number is given the symbol e and is the base of the natural logarithm. In a way, e is the counting base systems use whose rates of change depend on the size of their populations. If we determine the value of *factor(x)* for $x = 1000000$, the output is only 0.00005 % less than the 'exact' value of $e = 2.718281828459045$.

☑ Here, 'exact' means the value Mathcad has stored for e accurate to 15 decimal places.

Explorations

The Basics

1. For the function $f(x) = a \cdot b^{kx}$ given in the Warmup section, what is the effect of the increase in k on the behavior of the exponential function? What is the effect of variation in the constant a?

2. For $f(x) = a \cdot b^{kx}$, how would the plots appear if the value of k were negative? What would the effect of increasingly negative values of k be on the function's behavior?

3. Exponential decay and saturated growth (to a limit) occur so frequently in technological applications that it is worthwhile having some memorable guidelines to use. Develop tables for the following generic functions over the domain $x = 0$ to $x = 5$. The first function, $y(x) = e^{-x}$, describes exponential decay from an initial value of 1 to a final value of 0. The second, $y(x) = 1 - e^{-x}$, describes the growth from 0 to 1. The values in the decay table represent the % of the original amount left after a period x while those in the growth table represent the current % of the final saturated amount.

4. Re-express the exponent for the functions in #3 as $(-t/\tau)$. The denominator τ (tau) is termed the 'time constant' and is a measure of the time the system takes to decrease to e^{-1} of itself. Select a value of τ and plot the decay and growth relations as functions of time. What is the behavior of the decay and growth functions after a period of one time constant? after 5 time constants? If the value of the time constant changes, how does the appearance of the plots change?

5. The half-life of a substance is defined as the time taken for a sample to decay to 50% of its initial value. Determine the half-life of a system controlled by the following decay function: $amount(t) = 1000 \cdot e^{-t/10}$ with t measured in years and *amount* in kilograms.

6. Banks persistently advertise retirement investments which assure you will be a millionaire upon retirement. Determine the total value of your investment if you have deposited $1000 at the start of each of 40 years and are receiving 8% per year compounded semi-annually. The first $1000 has 40 years to grow while the second $1000 has 39 years and so on. The accumulation is paid out (or reinvested) at the start of the 41st year.

Beyond the Basics

1. Exponential functions exhibit rapid growth rates. Determine the point at which the function $y_1(x) = 1.1^x$ overtakes the power function $y_2(x) = x^{21} + 2$. Does this behavior apply to all exponential function versus power function situations?

2. The exponential function e^x can also be approximated by a summation given by

$$e^x = \sum_{n=0}^{\infty} \frac{x^n}{n!}$$. Determine the number of terms required for an estimate of e (or e^1)

accurate to 1 part in 10^{10}.

3. Inverse functions undo the effects of the function itself. As an example, for $y(x) = x^2$, x as a function of y is given by $x = \sqrt{y}$. Exchanging x with y yields the inverse function $y(x) = \sqrt{x}$. This final form undoes the output of the original function for $x \geq 0$. In a similar way, the inverse function of $y = 10^x$ is $y = \log(x)$ and of $y = e^x$ is $y = \ln(x)$. Plot each pair of functions on a separate graph. What characteristic of these inverse functions do you observe? Try other inverse function pairs to see if this is a universal characteristic.

3.4 Graphical Formats: Logarithmic

In an age of handheld computers, the use of counting tables (abacus) for calculation seems prehistoric. We assume that even though sophisticated electronic devices were obviously not around in the 1600s, people who had to calculate could with facility using the technology of the day. After all, it wasn't so long ago that calculation was done by hand or, at best, using mechanical calculators.

If we examine the process of adding two numbers together versus multiplying them by hand, the former process seems the simpler and less prone to error as the number of steps are less. In scientific calculations, where the number of steps can quickly mushroom, the relative strength of addition over multiplication would become apparent. It would even be reinforced if the counting device you had could only add and subtract. A method of using the device for multiplication and division would be close to a miracle and would save countless hours of tedium. Perhaps you can remember the first time an electronic calculator saved *you* hours or minutes of tedious number-crunching.

John Napier (Scotland, C16th) is widely held as the originator of logarithms. However, his work was preceded by others' work on powers of whole numbers and then later refined. Continental European mathematicians had actually developed similar systems as much as 15 years before but had failed to publish their work before Napier had.

Napier published the first consistent report on the use of powers applied to a fractional base as representations of numbers. Napier used a system based on powers of $10^7(1 - 10^7)$ or 999999. An Oxford professor, Henry Briggs, worked with Napier on refining the system. They agreed on a base of 10 for which $\log(10) = 1$ and $\log(1) = 0$. After Napier died in 1617, Briggs published a table of logarithms for numbers from 1 to 1000 accurate to 14 decimal places.

Warmup

Linear relations fit the scale of their rectangular Cartesian grid patterns pretty well. The axes, regularly divided, show the linearity or non-linearity of the function precisely over a finite range. Polynomial relations can be examined over various regions for characteristics of their behavior.

However, when the functions turn from power relations to exponential relations, linear axes can start to obscure the information. If the purpose of diagrams and plots is to convey huge amounts of information in a clear and compact form, the use of linear axes fails.

In Figure 1, the function $f(x) = 5^x$ is presented for $x = 0$ to 5. The table beside the plot shows the wide variation in the output for a relatively small variation in input.

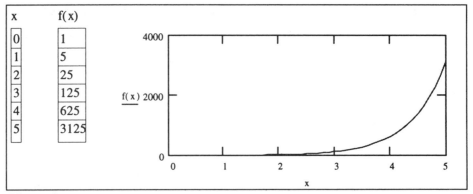

x	f(x)
0	1
1	5
2	25
3	125
4	625
5	3125

Figure 1

The first three data points are virtually swamped by the range in the output. A different plot type, the semi-logarithmic plot (one linear axis, one logarithmic axis) doesn't necessarily regain all the detail. However, the data points do become visible over the input range. Figure 2 shows the same function with the logarithm of the *y*-coordinates plotted versus *x*.

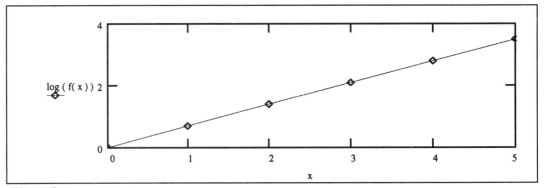

Figure 2

The application of the logarithm along one axis has seemingly changed the exponential function into a linear one. For the function $f(x) = 5^x$, the logarithm applied to both sides of the equation yields $\log(f(x)) = x \log(5)$. The application of the logarithm to the more general function $y = ab^x$ yields $\log(y) = x \log(b) + \log(a)$. In either case, the relation has a linear form, $y = mx + b$. The dependent variable *y* has been replaced by $\log(f(x))$. The slope *m* of the linear relation has been replaced by $\dfrac{\log[f(x_2)] - \log[f(x_1)]}{x_2 - x_1}$ for any two points x_1 and x_2 along the line. In the specific case of $f(x) = 5^x$, the slope is equal to $\log(5)$. The intercept of the vertical axis is given by $\log(a)$. For the function in Figure 2, the intercept is 0 or $\log(1)$ since the function can be interpreted as $f(x) = 1 \cdot 5^x$.

The semilog plot format is useful is showing large output ranges and also in evaluating whether or not data is of exponential or power function form.

In the above figure, the data points themselves have been lost along the *y*-axis. Unless you're incredibly quick at taking antilogs of numbers, the original points are obscured. Rather than plot the $\log(f(x))$, the plot region can simply be reformatted so that the *y*-axis is a logarithmic one. The scaling of the subdivisions contains the logarithm while the value of the data is retained. In the plot format dialog box, simply select the Y-Axis, Log Scale box. At the default plot size, you may only see the lines which represent the decades (powers of ten) or cycles. An increase in the plot size, as shown in Figure 3, will allow you to examine the detail of the logarithmic scale by introducing the secondary lines. However, these appear more clearly on the screen than on the printed page.

A characteristic of exponential functions is then their linearity when plotted on semi-log plots.

How would you go about defining the equation of this straight line? Once defined, how would you determine the exponential base? Care must be taken in remembering the vertical axis is now in a logarithmic form.

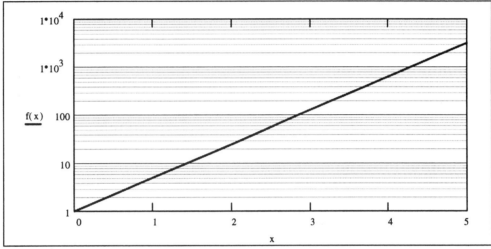

Figure 3

On the other hand, power functions of the form $f(x) = ax^b$, when plotted on semi-log plots, show a non-linear behavior. Application of the logarithm to each side of the equation yields a form of logarithm versus logarithm, $\log(f(x)) = b \log(x) + \log(a)$, with the slope of the straight line defined by the exponent $b = \dfrac{\log[f(x_2)] - \log[f(x_1)]}{\log(x_1) - \log(x_2)}$ for any x_1 and x_2 and with an intercept of $\log(a)$. Figure 4 shows the function $f(x) = x^4$ on a linear plot. Figures 5 and 6 show the same function represented on semilog and log-log plots. As before, the creation of a logarithmic axis is achieved by editing the plot format dialog box and, in the case of a log-log plot, selecting both axes.

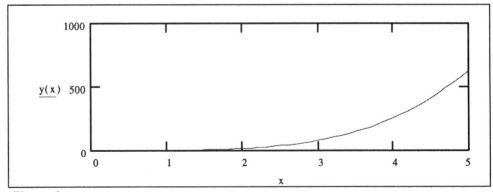

Figure 4

☑ With a log-log plot, the *y*-intercept of the linear relation can't strictly be defined since neither axis has a zero. Then $a = f(1)$ can be read directly from the graph.

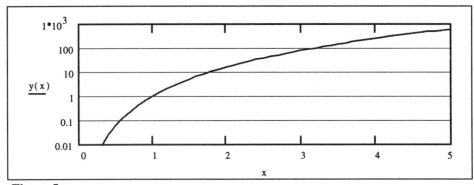

Figure 5

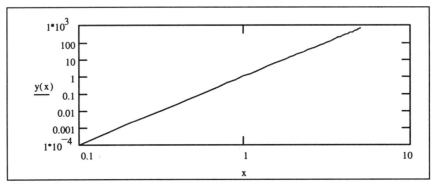

Figure 6

Explorations

The Basics

1. Plot the function $y(x) = 3 \cdot 5^x$ using i) linear, ii) semi-log and iii) log-log forms over the domain $x = 1$ to 100.
2. Plot the function $y(x) = 2\, x^5$ using i) linear, ii) semi-log and iii) log-log axes over the domain $x = 1$ to 1000.
3. Determine the power b in the function $f(x) = x^b$ for the following data set:

   (1, 1.0), (2, 11.3), (3, 46.8), (4, 128.0), (5, 279.5)
4. Determine the constant a and the base b in the exponential function $f(x) = a \cdot b^x$ for the following data set:

   (1, 2.72), (2, 7.39), (3, 20.09), (4, 54.60), (5, 148.41)

Beyond the Basics

1. For the data sets given below, one of the form $y(x) = a\, b^x$ and the other $y(x) = a\, x^b$, determine which is the exponential and which is the power function. As well, determine the values of the constants a and b for each form.

x	$y_1(x)$	$y_2(x)$
1	27	3.2
2	121.5	470.507
3	546.75	8718.123
4	2460.375	69180.216
5	11071.688	$3.449 \cdot 10^5$

3.5 Polynomial Functions

In a previous set of exercises, you have explored the creation of graph and table regions, the use of documentation and the effects of numerical and graphical formatting.

Here, we will be exploring the behavior of polynomials and the effects of degree and coefficient size.

In general, a polynomial function of degree n can be expressed as:

$$f(x) = a_0 + a_1 x + a_2 x^2 + a_3 x^3 + \cdots + a_n x^n .$$

The power is a nonnegative integer and the coefficients, a_n, are real numbers. The '0'th degree function is termed a constant function whereas the first and second degree polynomials are termed linear and quadratic respectively.

A function expressed as the quotient of two polynomial functions is described as a *rational* function. An example would be:

$$f(x) = \frac{3x^2 + 4x + 6}{x^2 + 2x + 1}$$

The general form of the polynomial can be expressed more compactly using the summation notation where the index covers the range from zero to the degree of the polynomial.

Mathcad offers a summation function Σ for $a_n x^n$ accessed by:
- typing $ a [n x ^ n <TAB> n, or
- choosing the summation icon from the menu on the left-hand side of the Mathcad window in Version 5.0, or
- by choosing from the Calculus palette in Version 6.0.

$$f(x) = \sum_{n=0}^{N} a_n x^n$$

Let's try both approaches now. In the first, the function will be explicitly stated with all coefficients clearly defined. In the second, the coefficients will be entered in the form of a vector, an ordered arrangement of $(N + 1)$ cells into which the coefficient amplitudes may be placed.

Warmup

First Approach

If we choose a quadratic, $y(x) = x^2 + 2x + 3$, we already have some sense of what the function will look like: a parabola opening upwards with its vertex at (-1, +2).

In order to generate Figure 1, a simple Mathcad file was created with the input variable domain and step size and the function $y(x)$ defined. The graphical format was adjusted so that grid lines are visible and the scales are numbered.

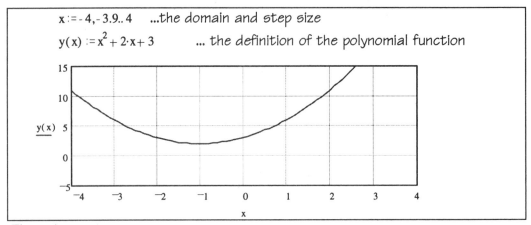

Figure 1

If we wanted to explore the behavior of a polynomial of higher degree or re-examine the same degree polynomial with different coefficients, we would have to edit $y(x)$ directly. This presents little problem as Figure 2 shows. You may wish to keep many files, each specific to a particular order and choice of coefficients. And this method is certainly the easiest way to explore at first.

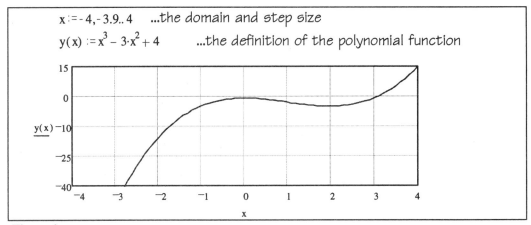

Figure 2

By editing the file you have created, plot a polynomial of degree 4. Choose any values of the coefficients and adjust the domain so that the behavior of the function is evident. Save the file under a different name.

Now, re-edit the file for a polynomial of degree 7. Or choose not to. There is quite a bit of typing involved as the degree increases. Which brings us to the second approach.

Second Approach

Here let's try to structure the polynomial plotter file so that it is more easily editable and scaleable, so that our exploration is not impeded by the editing process. As well, this approach uses vectors, columns of numbers defined by the position they hold within the array. For our application, the numbers represent the coefficients of each degree.

A vector a_n of coefficients is created by defining an index n over the whole range, up to and including the highest degree (n : = 0 .. N) and then by typing "a [n : a_0, a_1, a_2, ... a_N".
Here a_0, a_1 ... represent the *actual numerical values* of the coefficients. You will have created a column of $(N + 1)$ numbers each defined by its respective position within the column. Any of

these numbers or elements can be extracted from the vector. For example, the value of the $n = 3$ element is given by typing "n [3 = ".

With the layout of Figure 3, the degree can be adjusted easily by editing the definition of *N*. Coefficients can likewise be added or deleted as needed. The output of this particular polynomial is given in Figure 4.

Try your own version of this file for a number of different degree polynomials. Be aware of the scale of your graph as the choice of coefficients may drive the function out of the previously defined plot region.

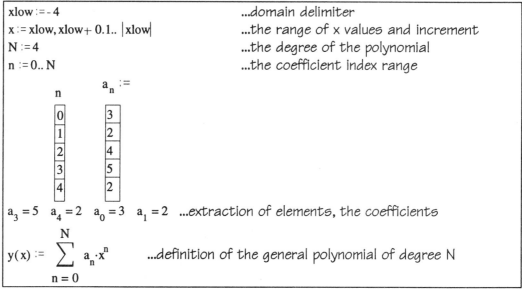

Figure 3

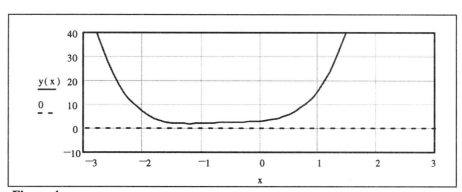

Figure 4

Explorations

The Basics

1. Plot the function $y(x) = 3x^2 + 2x + 6$ for a suitable domain and step size. Determine the effect of each of the coefficients and the constant on the quadratic function by altering each in turn using either your own values or the ones suggested below:
   a) $y(x) = -3x^2 + 2x + 6$
   b) $y(x) = 5x^2 + 2x + 6$
   c) $y(x) = 3x^2 + 2x - 6$
   d) $y(x) = 3x^2 + 5x + 6$
   e) $y(x) = 3x^2 - 3x + 6$

2. Based on your analysis of #1, predict the general shape, vertex location and y-intercept of the function, $y(x) = -4x^2 + 6x - 3$. Show the validity of your prediction using tables and plots.

3. Create a coefficient catalogue. Choose a polynomial of low order ($N < 5$) and systematically alter the values of only one of the coefficients. Examine and observe its effect on the behavior of the function.

4. Examine the behavior of different degree polynomials for the domains $|x| < 1$ and $|x| > 1$.

5. Even functions are functions for which $f(x) = f(-x)$. Whereas, an odd function is one for which $f(x) = -f(-x)$. Examine the symmetries of polynomials containing pure odd powers, pure even powers and mixtures of even and odd powers. Note that the definitions of odd and even apply to the standard coordinate axes ($x = 0$ and $y = 0$) and the origin. Functions may display symmetry with respect to other lines.

Beyond the Basics

1. Use the plot of a high degree polynomial to provide estimates of the initialization points required by the Root function or Solve Block and solve the polynomial equation $f(x) = 0$.

2. Estimate the function which best approximates the following data points: (-5,441), (-2,0), (0,16), (2,0), (4,144), (5,441).

3. Create a rational function and plot it. Load the Symbolic Processor and use it to factor each of the numerator and denominator. Depending on the coefficients you have selected, you may not be able to obtain any simple (whole number type) factors. The expression $x^2 + 5x + 6$ would factor to $(x + 3)(x + 2)$ whereas the expression $x^2 + 4x + 6$ would simply be returned unchanged. Rather, use the 'Solve for Variable' option under the 'Symbolic' menu. This option solves the equation, $f(x) = 0$. For the quadratic solution set of x_1 and x_2, the original expression factors into $(x - x_1)(x - x_2)$. What do the zeros of the numerator and the zeros of the denominator (i.e., those values of x which drive the expression to zero) tell you about the behavior of the plotted function.

4. Create a menu driven tutorial program. Let the degree of the polynomial and the values of the coefficients be easily editable and allow space for cutting, pasting and observations. Conclude the file with a series of questions about the nature of polynomials.

3.6 Graphing Singularities and Discontinuities

So far the functions you have plotted have been pretty well behaved. They have displayed no sudden leaps, no jaggedness and no undefined points. You chose a suitable domain, a convenient step-size, and asked Mathcad to plot the function $f(x)$ over the input range. The major difficulties you may have encountered were in selecting the plot region limits appropriately and choosing between all the various display options (colors, plot types, show markers, grid lines, autoscale, and so on).

All that will change with this set of explorations. We will examine the plotting of functions whose character is problematic, of functions that blow up to infinity at certain values of x and of functions displaying vertical steps and pointed vertices.

Warmup

Singularities

A function $f(x) = \dfrac{1}{x}$ or any function with negative powers in x present difficulties when you attempt to evaluate or plot them. At some point, the denominator is driven to zero and, since division by zero is undefined, the function is undefined at this point. A singularity exists at that particular value of x.

Mathcad offers a variety of ways of handling this difficulty. However, we must be aware of the problem of the singularity so that we understand any unexpected output. Plots are a powerful way of conveying information. If the process obscures the information, the process works against us.

A plot over a small range including the origin would show a break at $x = 0$ for the function above. If we define the range as $x := -3 .. 3$ and then try to tabulate the results, Mathcad freezes and returns an error message.

$$x := -3 .. 3 \qquad\qquad f(x) := \frac{1}{x} \qquad f(x) = \boxed{\text{singularity}}$$

Figure 1

Obviously, the domain includes $x = 0$ and division by zero causes the message to appear. Mathcad does however display some peculiarities here when we redefine the range using a smaller step-size. If we let $x := -3, -2.9 .. 3$ then a table of values would show the output for $f(x)$ at $x = 0$ to be a number on the order of $4.5 \cdot 10^{15}$. Try to evaluate $f(0)$. You probably received a singularity message again. What is going on?

Try restricting the range to $x := -0.2, -0.1 .. 0.2$. Another error message. Mathcad's definition of zero seems to skip around depending on the size of the range. In the above example, the value of x evaluated for was on the order of 10^{-15}, close to zero but not zero. You can examine this behavior by changing the zero tolerance in the global formatting and see the effect your changes have on Mathcad's behavior.

If we now wish to represent this function, with its singularity region included, the plot (Figure 2) shows a sharp spike at the singularity. With the limit selection left to Mathcad, the function shows zero everywhere but at the origin. Not a very informative plot, overall.

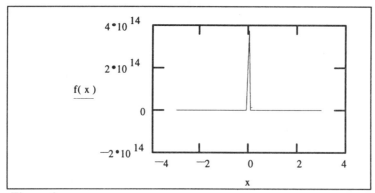

Figure 2

Now, let's restrict the plot limits along the *y*-axis by selecting the plot region and altering the vertical axis limits.

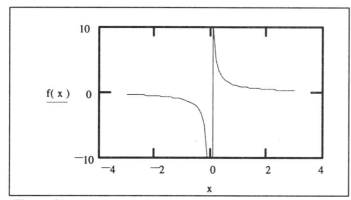

Figure 3

This plot (Figure 3) is a better indication of the function's behavior near the origin. We see the progression to minimal output as |x| grows very large and, conversely, the move to infinite output near the origin. However there appears to be a vertical line, an attempt to connect negative infinity to positive infinity. How do we get rid of this artificial line? Try changing the step-size definition of the domain. You may end up creating a better defined plot with a more vertical line.

The problem may be that our domain definitions have all included the singularity input within their definitions. If we simply skip over the problem by using a wiser choice of domain, will the problem disappear?

Perhaps we need to break the domain into separate regions, one region for each section of the problem and overlay them on the same plot area. In this way, no detail is obscured at the singularity.

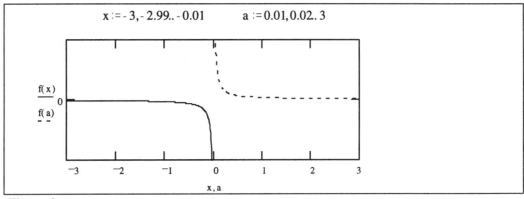

Figure 4

Figure 4 shows the function $f(x)$ split into two separate domains. The vertical line is gone and the information is clear and distinctive.

Discontinuous Functions

Functions may also exhibit discontinuous behavior over certain ranges. An otherwise continuous function may exhibit a step like behavior or a cusp or vertex at a point x. An example of a discontinuous function is

$$f(x) = \begin{cases} x - 3 \text{ for } x \le 0 \\ -4x + 2 \text{ for } x > 0 \end{cases}$$

To plot this we could define two separate functions, one for each continuous part of the function. Or, we could use the **if** utility Mathcad offers. The syntax of the if statement is:

if(conditional statement, truevalue, falsevalue).

It provides a more elegant way of displaying this type of function as shown in Figure 5.

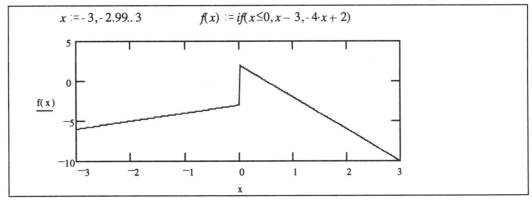

Figure 5

☑ The conditional statement 'Less than or equal to' is typed using CTRL-9. Control key combinations for other conditional statements can be found in the Help documentation.

Heaviside Step Function

Another tool useful in creating discontinuous function (as you may need when a signal is suddenly turned on) is a step function defined by Oliver Heaviside, a British mathematician and physicist we will meet again when discussing differential equations and transforms.

The Heaviside step function is defined as: $\Phi(x) = \begin{cases} 0 \text{ for } x < 0 \\ 1 \text{ for } x \geq 0 \end{cases}$

You could define this yourself using the *if* function as: Heaviside(x) : = if (x < 0, 0, 1)

but Mathcad provides the built-in $\Phi(x)$ function. Figure 6 shows both the built-in and the custom built functions.

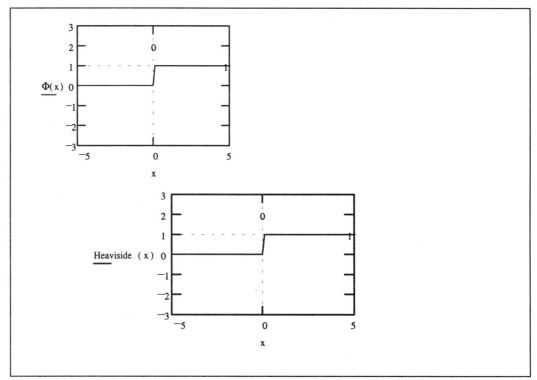

Figure 6

| Explorations |

☑ The conditional statements (=, <, >, ...) can be combined using AND and OR logical operations. The AND operator takes the form (conditionA)·(conditionB) while the OR operator is represented as (conditionA)+(conditionB). For example, a 1 (or ON state) would be returned only if both (conditionA)·(conditionB) were true while a 1 would be returned if either or both of (ConditionA) + (ConditionB) were true. Otherwise, a zero (or OFF state) is returned to the function.

The Basics

1. Plot each of the two functions given below. The domain should include all regions of singularity.

   a) $f(x) = \dfrac{1}{x^2 - 4}$　　　　　　b) $f(x) = \dfrac{x+4}{x^2 - x - 6}$

2. Use the Heaviside Step Function to turn the function $f(x) = 5x - x^2$ ON at $x = 4$. To the left of this point, the function will output zero. And to the right of this point, the function will present a quadratic. Try to make the transition as sharp as possible.

3. Switch the function $y = \sin(t)$ on at $t = 3.0$. The sine function presents a smoothly varying wave over all t. How would you go about turning the function OFF an interval $\Delta t = 10.0$ later? This ON-OFF pattern produces a pulsed wave. Try setting the limits to the beginning and end of the waveforms (i.e., points where $y(t) = 0$).

4. Using the conditional statements and operators from #3, create a square wave pulse of duration $\Delta x = 1.5$ and height $\Delta y = 4$ centered at $x = 2.5$.

Beyond the Basics

1. Develop a routine which accepts two functions $f_1(x)$ and $f_2(x)$, determines their point of intersection (if any) then connects them into a discontinuous function at that point (x-intercept, y-intercept). As a start, try $f_1(x) = 4x + 5$ and $f_2(x) = -3x^2 + 4$ with $f_1(x)$ for $x < (x$-intercept) and $f_2(x)$ for $x > (x$-intercept). Note: to determine the point of intersection, the 'Solve Block' routine can be used.

2. Using repeated conditional statements, create a train of square pulses of equal duration and equal height separated by a common interval.

3. The Heaviside function represents a one step transition. Use the vector index techniques developed in Section 3.5, 'Polynomial Functions', to create a staircase effect. For k : = 0 .. 10, let $x_k := k$ while $y_k := k + 1$. On the plot, select the Step trace type for the graph of these vectors. Once this is set, try to create another ascending staircase with $\Delta x = 2$ and $\Delta y = 3$. How would you create a descending pattern?

3.7 Conic Sections

The study of the intersection of planes with a 3-dimensional cone has extended from the time of the ancient Greeks (circa 350 B.C.) to the present. These sections are still used in defining the shape of gears, cams, satellite dishes, light reflectors, suspension cables and bridge arches.

When a plane intersects a cone it can generate four different intersection patterns: a circle, an ellipse, a parabola, or a hyperbola. The study of these shapes, their translations and transformations makes up a substantial section of the study of Plane Geometry.

Warmup

We will be examining the conic sections in their rectangular forms, as expressions of *x* and *y* in a 2-dimensional plane. In a later section, the use of polar and parametric relations will be explored. These more elegant forms greatly reduce the complexity of representing the conic sections.

A 3-dimensional cone can be generated in Mathcad by defining a series of circles parallel to the *x-y* plane. As the circle moves upward along an axis, the radius decreases in a linear fashion from its maximum value on the *x-y* plane to zero at the vertex.

Figure 1 shows the setup of variables needed to create the cone surface plot. The Surface Plot is defined as the height of the surface at each grid point in the *x-y* plane. The **if** conditional function was used to set the output at zero if the radius of the circle extended beyond defined limits. The radius of the base is 2.0 units and the height is 5.0 units.

The result is the representation of a matrix whose elements are the heights of the surface. Figure 2 shows both the surface plot and a contour plot for the cone.

$$N := 30$$

$i := 0..N$	$j := 0..N$	vector index
$a := 2$	$b := 5$	cone slope parameters
$x_i := -3.0 + i \cdot 0.2$	$y_j := -3.0 + j \cdot 0.2$	definition of x, y grid points
$h(x,y) := -\dfrac{b}{a} \cdot \sqrt{x^2 + y^2} + b$		equation of straight line envelope for cone
$z(x,y) := \text{if}\left[\left(x^2 + y^2\right) > a^2, 0, h(x,y)\right]$		definition of height above x-y plane
$CONE_{i,j} := z\left(x_i, y_j\right)$		definition of conic matrix

Figure 1

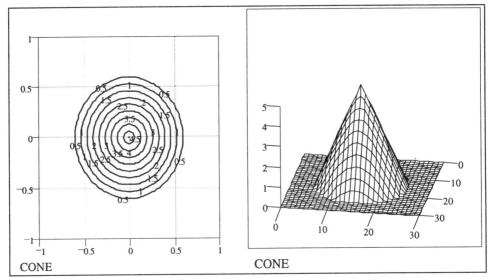

Figure 2

The Circle

Probably one of the most satisfying geometric shapes is the circle. Symmetric, harmonious, perfect, pure, the circle has fascinated generations of mystics and engineers.

The circle was originally held to be of divine origin. The heavens above and the stars within moved in great circles while on earth humans and their products moved in straight lines. Circles were a representation of the eternal. Humans, being frail and mortal, had only their lines to content themselves with. Our interpretations have changed over the years but the usefulness of the circle continues.

As a conic section, a circle represents the intersection of the cone with a plane parallel to the *x-y* plane. As the plane travels upward from the base of the cone to the vertex, the circle goes from its maximum radius to extinction.

Figure 3 includes the definition of the truncated cone using the information from Figure 1. Figure 4 shows the cone from Figure 2 terminated at a height of 2 units. Beside it, a contour plot shows that the plateau at the top of the truncated cone is a circle. The boxiness is a result of the matrix pattern Mathcad uses for Surface Plots. The pattern was created using nested conditional **if** statements and will be examined in the Explorations.

$$zz := 2 \qquad \text{truncation limit}$$

$$z(x,y) := \text{if}\left[\left(x^2 + y^2 \right) > a^2, 0, \left(\text{if}(h(x,y) < zz, h(x,y), zz) \right) \right] \qquad \text{truncated cone definition}$$

$$CONE_{i,j} := z\left(x_i, y_j \right) \qquad \text{matrix definition}$$

Figure 3

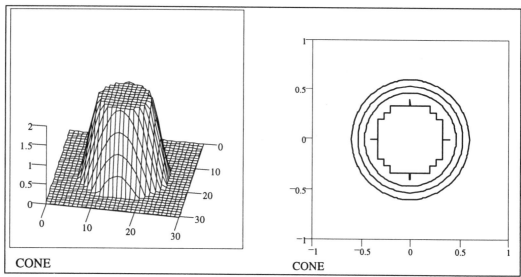

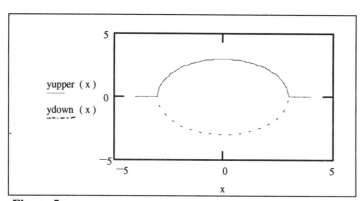

Figure 4

The points along a circle are equidistant from its center. In general, a circle's points have to satisfy the relation $(x - h)^2 + (y - k)^2 = r^2$ where (h, k) represents the center point and r represents the radius.

To plot the circle in Mathcad we need to be aware of the difference between a function and a relation. Functions are one-to-one, one output for every one input. However, a circle plotted on a graph displays two outputs for every input. In Mathcad, we can *circle* the problem by generating two separate functions, $y(x) = \sqrt{r^2 - x^2}$ and $y(x) = -\sqrt{r^2 - x^2}$, the former for the upper half of the circle (here, centered at the origin) and the latter for the lower half. However, this solution in turn creates its own set of problems.

Figure 5

The initial plot of the circle centered at the origin, as shown in Figure 5, is full of obscuring designs: the circle appears compressed along the vertical direction, the plot types (and colors) are different top and bottom, and the bottom plot seems to have left some 'wings' along the *x*-axis.

The plot types can easily enough be made the same type and color by setting the X-Y Plot Format selections. The scaling problem can be adjusted by altering the size of the plot region or by redefining the axes limits. As for the wings, the domain can be redefined. The **if** function would not be much help here as it would output a zero and this 'solution' would show up as points along the *x*-axis.

Once we have cleaned up the circle, it appears as in Figure 6, a perfectly acceptable sight.

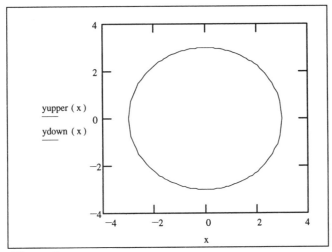

Figure 6

The Ellipse

If the intersecting plane is tilted slightly from being parallel to the base, the radial symmetry of the circle is broken. The intersection curve seems to be compressed along one of its directions. In fact, the circle is one particular case of this more general shape.

For an ellipse centered at the origin, the equation is $\dfrac{x^2}{a^2} + \dfrac{y^2}{b^2} = 1$ where *a* and *b* are the lengths of the semi-major and semi-minor axes respectively and $a > b$ by convention. As in the case of the circle, the relation has to be broken into an upper and a lower function to be plotted.

The focus points for this ellipse would occur at (+*c*, 0) and (-*c*, 0) where $a^2 - c^2 = b^2$. Figure 7 shows an ellipse centered at the origin with $a = 5$ and $b = 2$. The foci appear at (+4.583, 0) and (-4.583, 0).

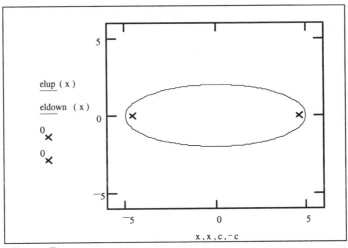

Figure 7

The Parabola

As the intersecting plane is tilted even more, there comes a point at which one side of the plane cuts through the base of the cone rather than its side. A parabolic shape emerges.

On the *x-y* plane, the parabola can be defined by its vertex, focal point and directrix. The directrix is a straight line used as a reference point for the generation of the parabola. In a physical application, the focus represents the point to which rays of light or sound parallel to the *y*-axis would be focused.

The line of the parabola is generated by drawing points equidistant from the focal point and along a perpendicular from the directrix. For a parabola symmetric to the *y*-axis and with its vertex at the origin, the equation is $x^2 = 4py$. The focus is located at $(0, p)$ and the directrix is defined by $y = -p$.

Figure 8 illustrates this for the parabola $y = 3x^2$. The vertex $(0, 0)$ is midway between the focus at $(0.083, 0)$ and the directrix defined by $y = -0.83$.

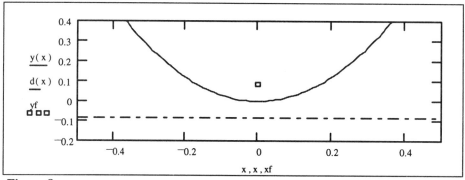

Figure 8

In most cases of plotting general parabolas, $y = ax^2 + bx + c$, the details of the directrix and focus are implicit and hardly noticed. However, when it comes to designing a parabola with certain characteristics, knowledge of these is essential.

The Hyperbola

At first glance, the hyperbolic equation looks much like the elliptical equation. However, the hyperbolic shape is created when the intersecting plane is steep enough to cut through both a bottom and top cone joined at the vertices.

Whereas the ellipse is generated by keeping the sum of distances from the focal points constant, the hyperbola uses the difference of the distances as its constant. For a hyperbola, $\dfrac{x^2}{a^2} - \dfrac{y^2}{b^2} = 1$. The focal points $(+c, 0)$ and $(-c, 0)$ are defined by $a^2 + b^2 = c^2$.

Figure 9 shows a hyperbola with $a = 4$ and $b = 3$. The foci are $(+5, 0)$ and $(-5, 0)$ while the vertices are $(+4, 0)$ and $(-4, 0)$. Notice again that the pattern needs to be split into an upper and lower function and that the plot types need to be the same. Special attention needs to taken with the definition of the domain. Otherwise, lines appear where the function is undefined.

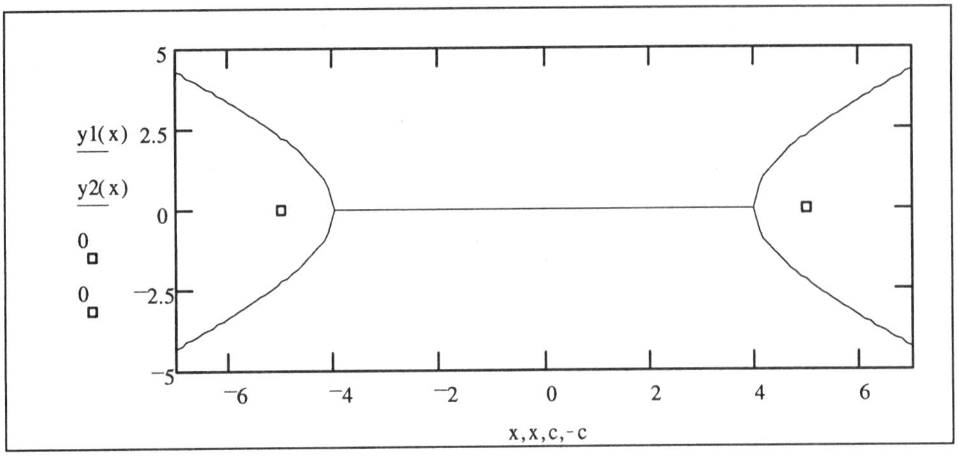

Figure 9

In the case of this hyperbola, how would you avoid the line joining the vertices?

Explorations

The Basics

1. How do you translate the center of the circle to any point on the plane? Create a plot of four circles of different radii, each of which is centered on a different point. Pay attention to your scaling.
2. Would the same technique apply for the translation of vertex of a parabola to any point on the plane? Show the same parabola with its vertex translated to three different positions on the plane.
3. Create a plot with three circles of different radii whose diameters fall along a line. Now, move the circles so that they are just touching (as one gear would another). Finally, move one of the touching circles off the line of diameters (to a maximum of 90°).
4. Design a parabolic dish antenna with a focal point 30 cm above its vertex and a circumference at the open face of 9.43 meters.
5. For an elliptical tank, determine four different configurations of *a, b* and length *L* all of which hold the same volume *V*. The area of an ellipse is πab. Would one configuration be more practical than the others?

Beyond the Basics

1. In the *Basics* exercise, we have moved the vertices of parabolas and the centers of circles and ellipses around on the *x-y* plane. However, the orientation of the form has not changed. How would you go about taking one of these conic sections and rotating it around the origin by a fixed angle? See Appendix 1 for a description of axis rotation. It all depends on your point of view.
2. For the hyperbola plotted in the Warmup section, examine its behavior over a much larger domain. What do you notice as the absolute value of *x* gets large? These lines, the asymptotes, can be defined by expressing the $y(x) = \pm b\sqrt{\dfrac{x^2}{a^2} - 1}$ relation for $x \gg a$. In the wings, the ratio of *a* to *x* becomes infinitely small.

3. The cone in Figure 2 was plotted by defining the height as

$$z(x,y) = \frac{-b}{a} \cdot \sqrt{x^2 + y^2}$$ with a being the radius of the cone's base and b the height of the cone above the x-y plane. Mathcad's **if** conditional statement was used to drive any output beyond the maximum radius to zero. In Figure 4, a second conditional statement (on the height z) was placed within the first to create the truncated cone. Use this technique to create an elliptical and parabolic section.

4. Bridge arches use conic sections due to the efficient way stresses are distributed. Create an under-bridge support arch using a single circular, parabolic, elliptical or hyperbolic section. The bridge spans 100 m and its surface is 30 m above the floor of the hollow.

3.8 Parametric Curves and Polar Coordinates

The graphing and plotting you have done so far has been primarily of functions in the rectangular coordinate system. You have examined linear and power functions, exponential and logarithmic plots.

In this set of exercises we will extend our knowledge of graph types and of Mathcad's ability to plot parametric and polar curves.

Circles, ellipses, and other closed or open figures in which a value of the input variable is associated with two or more values of the output variable are not functions. The definition of a function is specific in allowing only one value of y for each value of *x*. Mathcad cannot treat these relations using the functional form.

In the earlier versions of Mathcad, parametric curve plotting was included within the available tools while polar curve plotting had to be emulated using the rectangular to polar coordinate transformations. Polar curves had to be tricked out of the software using the parametric form.

Parametric Curves

A parametric curve is one in which the coordinates *y* and *x* are the dependent variables associated with an independent and invisible variable *t*. As *t* varies over its range y and x take on values specific to the relations $y = y(t)$ and $x = x(t)$. The points *x* and *y* for a given *t* are then plotted against one another on a rectangular or logarithmic grid.

The parametric representation can easily be confused with the functional representation $y = y(x)$. The graphs created by each process may even look a lot alike.

In Figure 1(a) below, the function $x(t) = t^2$ is plotted. This represents the *x* versus *t* relation for one part of the parametric curve given in Figure 2. Both figures appear to represent functions. However, the arguments in Figure 1(b) are each dependent on a hidden variable *t*.

Figure 1(b) below displays the parametric relation: for $t = 0$ to 10, $x(t) = t^2$ and $y(t) = 3t - 4$.

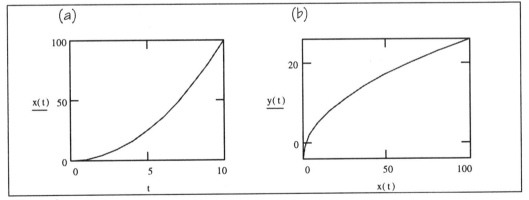

Figure 1

Polar Curves

A 2-dimensional polar graph is a plot of the relationship between the radius from the origin and an associated angle. The function is expressed as $r = r(\theta)$. A circle of radius 4.0 units can be represented by $r = 4$ over the range of angles 0 to 2π while a four-leafed rose can be plotted using $r = a \sin 2\theta$.

While the older versions of Mathcad allowed you to plot polar graphs by turning them into parametric curves using the rectangular to polar transformations ($x = r \cos \theta$, $y = r \sin \theta$), the newer versions contain polar plotting routines which are simpler to use and which create more distinctive polar plot regions. See Figure 2 below.

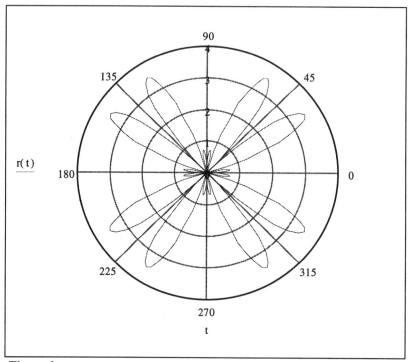

Figure 2

In order to plot these curves, the variable used is the angle θ in radians. This variable is not restricted to the range 0 to 2π or from 0° to 360°. However, in the case of a some closed curves, points beyond this range simply overlap previously drawn points. For an open curve, an extended range allows a complete analysis of the behavior of the curve.

☑ Mathcad seems to exhibit a quirk here as its default angle format is in radians yet the polar plot grid is in degrees. This problem disappears with continued use and familiarity!

Warmup

Parametric Curves

Let's examine the one-dimensional motion of a mass in a gravitational field, a ball thrown straight up. The equation for the position of the mass is $x(t) = x_0 + v_0 t - 0.5 \, g t^2$ where x_0 is the initial position, v_0 is the initial velocity and g is the acceleration due to gravity. The equation for the velocity of the object as a function of time is $v(t) = v_0 - g t$.

Each of these relations can be plotted on its own by choosing a suitable range for t and values for x_0, v_0 and g. The plots of x versus t and v versus t are given below (in Figures 3 and 4, respectively) for t from 0 to 7 s and $x_0 = 2$ m, $v_0 = 30$ m/s and $g = 9.8$ m/s^2 .

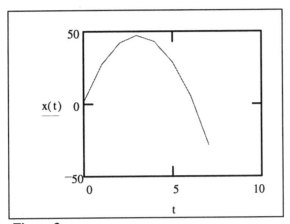

Figure 3

Notice that while the mass started from a height of 2.0 meters above the ground, it seems to have penetrated below the ground's surface. The equation of motion needs to be complemented by some common sense and an analysis of the time at which the ball returns to the surface. The same reasoning applies for the velocity graph below.

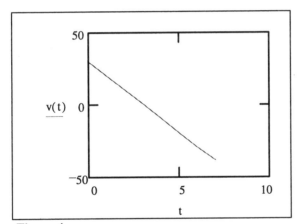

Figure 4

The parametric form can now be used to explore the phase space diagram of the motion, the time progression of the position versus the velocity for this system (Figure 5).

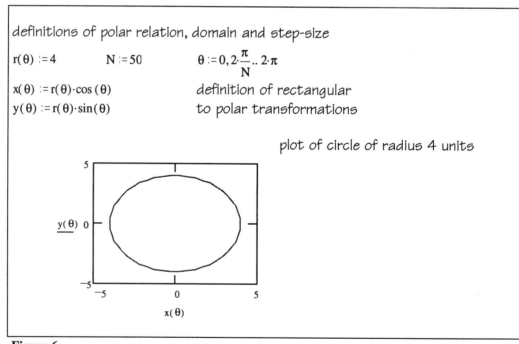

Figure 5

Old Polar Form

To display a polar relation, the parametric form can be used with the angle θ as the parameter for points within the *x-y* plane. The polar relation is expressed as $r = r(\theta)$ and the polar to rectangular transformations, $x(\theta) = r(\theta)\cos(\theta)$ and $y(\theta) = r(\theta)\sin(\theta)$ are included directly afterward. The parameter is defined so that the full pattern of the plot is represented. The plot of a circle of radius 4 units would involve the clear definition of all these relations. See Figure 6 below.

definitions of polar relation, domain and step-size

$$r(\theta) := 4 \qquad N := 50 \qquad \theta := 0, 2 \cdot \frac{\pi}{N} .. 2 \cdot \pi$$

$$x(\theta) := r(\theta) \cdot \cos(\theta)$$
$$y(\theta) := r(\theta) \cdot \sin(\theta)$$

definition of rectangular
to polar transformations

plot of circle of radius 4 units

Figure 6

Notice that the default scaling of the plot makes the circle look like an ellipse. If you examine the *x*- and *y*-scales, this "error" is easily corrected by changing the size of the plot until the scaling is equal along both axes.

More elaborate designs may be created by editing the expression for $r(\theta)$. Something has to be left for the Explorations!

New Polar Form

The introduction included a plot which was created using Mathcad's Create Polar Plot option from the Graphics menu (shortcut key CTRL-7) or by selecting the Polar Plot icon from the palette (in Version 6.0).

Figure 7 shows the functional relationship for the polar plot already given in Figure 2. A range for θ can be defined from 0 to 2π radians. Or, if you prefer to work in degrees, the angle can be multiplied by the unit 'deg'. A plot of the function is given. Further analysis of these patterns is left to the Explorations.

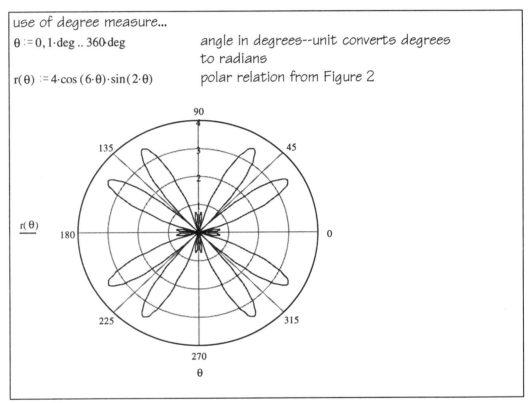

Figure 7

A polar plot format dialogue box can be accessed by double-clicking on the plot region. Grid lines and plot types can be changed according to your preferences.

Explorations

Choose any of the various outlined plotting routines to examine the explorations. Your choice may depend on the version of Mathcad you are running or the ease of use of one routine over another.

When describing circles and other forms with rotational symmetry, pay attention to the scaling of the plots.

If any of the explorations seem like a LOT OF WORK, you are probably making it too hard on yourself by not creating a *template* file and adjusting the variables in the template for each new plot type.

The Basics

1. For the Warmup example of the ball thrown vertically into the air, determine the total time of flight from its takeoff 2.0 m above the ground to impact. Use this information to plot the phase-plot for the duration of flight.
2. Use the Old Polar and the New Polar techniques to plot 4 concentric circles.
3. The following polar forms create multi-leafed patterns. Explore their behavior as the values of the constants *A, B* and *C* are changed.

   a) $r = A \cos (B\theta)$ b) $r = A \sin (B\theta)$ c) $r = A \cos (B\theta) \sin(C\theta)$
4. Parametric curves have been studied over the ages and many have acquired fanciful names. Here are just a few whose behavior you can examine as the constants are varied.

   a) $x = A \cos^3(\theta)$ and $y = B \sin^3(\theta)$ (the evolute of the ellipse)

   b) $x = A \cot(\theta)$ and $y = A \sin^2(\theta)$ (the Witch of Agnesi)
5. A common microphone range and antenna pattern is the Cardioid (heart-shaped) given by the relation $r(\theta) = a\, [\, \cos(\theta) + 1\,]$ where *a* is a constant. Examine the effect on the pattern of size and sign of *a*. What would the effect be if the angle θ were replaced by $n\theta$ where *n* is greater than 1?
6. Lissajous patterns can be created on an oscilloscope by feeding separate wave generator signals into each of the channels. You can create your own by examining the parametric relations $x(t) = A \sin(Bt)$ and $y(t) = C \cos(Dt)$ for any choice of the constants *A, B, C* and *D*.
7. Examine the behavior and domain of the following open curves and spirals. The angular measure θ is in radians.

   a) $r(\theta) = A\theta$ (the Spiral of Archimedes)

   b) $r(\theta) = A/\theta$ (the Hyperbolic Spiral)

   c) $r(\theta) = e^{A\theta}$ (the Logarithmic Spiral)

Beyond the Basics

1. Projectile motion is usually analyzed as motion in two dimensions, vertical height (*y*) and horizontal distance (*x*) from the projection point. The relations which describe the motion of a projectile launched from a position (x_0, y_0) at an angle θ with respect to level ground with a initial velocity v_0 are:

   $x(t) = x_0 + v_0 \cos(\theta)\, t$

   $y(t) = y_0 + v_0 \sin(\theta)\, t - 0.5gt^2$

   Air resistance has been ignored in these equations. Examine the behavior of the projectile for different values of the constants. How does the value of *g* affect the time of flight and the range? Is there an angle θ for which the range reaches a maximum?
2. The relation $r(\theta) = \dfrac{a}{b + c \sin(\theta)}$ can generate the conic sections depending on the choice of the constants *a, b* and *c*. Determine by experiment the conditions necessary for the creation of each of the sections.
3. How would you translate any of the polar or parametric plots you have created from their center at the origin to a new origin located at (x, y)?
4. How would you rotate any of the patterns you have created around the origin by a fixed angle ϕ ?

Chapter 4: Systems of Linear Equations

The functions and plots of Chapter 3 can be viewed as recipes for the transformation of some input variable into its final form as an output variable. Another interpretation is that of a mapping of one set of numbers onto another set with the functional relation providing the linking process.

The list of similar interpretations could grow long. For this chapter, linear equations are seen as constraints, sets of well defined rules which differentiate the included sets of numbers (those which satisfy the condition) from the excluded sets.

In this way, the infinite number of points generated from one linear relation in two dimensions can be collapsed to a point given the application of a second constraining relation to the output of the first.

The number of connected equations a system can generate is an indication of its flexibility, its variables and degrees of freedom. In these exercises, we will examine various techniques which can be applied to the solution of these systems. Each technique has its strengths and weaknesses but each is valid as it reveals a different aspect of the 'same' linear system.

4.1 Graphical and Numerical Solution Techniques

The two-dimensional functions we have examined so far create single curves on a flat plane. The points along the curve are those which satisfy the condition or constraint set out by the functional relation. However, if more than one constraint has to be taken into account at a time, then the variables allowed from the first equation are again selected for based on their appropriateness within the second.

A simple engineering problem involving a span across two supports creates a relation in the sum of the forces and another in the torques. Together these provide the necessary stress analysis of the static structure. One relation is simply not enough of a condition as it provides for an infinity of solutions. In two dimensional cases, the application of the second set collapses this infinity to a single point if a solution exists.

In basic electronics, the equations governing the response (current) of a DC system to an applied force (voltage) can generate a high dimension system for the simplest configuration of elements. As the circuit includes more current loops the number of equations increases dramatically.

In the world of high finance, competing stresses on the growth of money create systems whose outputs are inclusive of the conditions placed on the variables.

Although there exist programmable graphics calculators which can solve high dimension systems in the click of an ENTER button, the ability to edit and contrast numerical and visual information is one of the advantages of using Mathcad to solve these system problems.

In this set of explorations, we will be examining the use of visual and numerical techniques in the solution of systems, starting with a basic 2x2 system (2 equations in 2 unknowns) and working up to larger dimension systems. Although the visual approach can provide highly accurate information, the numeric approach involves an iteration process which is faster and more reliable given the same tolerance demands for both techniques.

The more analytical techniques will be explored in further sections.

Warmup

A simple 2-dimensional system of linear equations can be represented as

$$1) \quad y(x) = m_1 x + b_1$$
$$2) \quad y(x) = m_2 x + b_2$$

where m represents the slope and b, the y-intercept. While this form allows quick plotting, it may not be the most practical form for applying analytical routines to the system.

By itself, the first relation (1) offers an infinite number of possible values of x and y. These coordinate pairs are constrained by the relation like beads on a wire. The second relation, by itself, follows the same behavior. If both relations are true simultaneously, the infinity of points each has collapses to a single common point of intersection. However, the relations may also represent parallel lines with no solution or even the same line with an infinite but trivial solution.

Many of the analytic solution techniques used to solve systems of equations assume an answer already exists as variables from one equation are fed directly into the other.

For our first system, the values of (x, y) have to satisfy the relations:

$$3x + 7y = 90 \text{ and } 7x - 2y = 25.$$

In slope-intercept form, these equations transform to:

$$y = \frac{90 - 3x}{7} \text{ and } y = \frac{7x - 25}{2}.$$

Figure 1 shows that the solution to the system occurs in the region of the point (6.0, 10.0). The plot also shows that a solution exists. The estimate of the intersection point using a picture may be as precise an identification as you need.

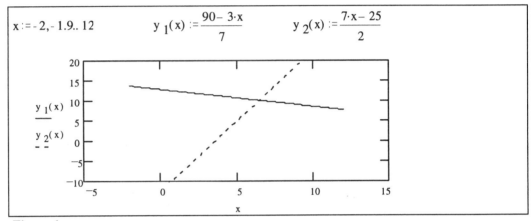

Figure 1

Or, the Solve Block routine can be used to determine a more accurate value of the solution. The visual presentation can be used to provide good estimates of the initialization points in x and y. Figure 2 shows the use of Solve Blocks applied to the same system. The solution is output as a vector whose elements are the coordinates x and y of the point of intersection.

$$x := 6 \quad y := 10 \quad \text{...estimates} \quad TOL := 0.001 \quad \text{...tolerance setting}$$

$$\text{Given} \qquad 3 \cdot x + 7 \cdot y = 90$$

$$7 \cdot x - 2 \cdot y = 25$$

$$\text{SolutionA} := \text{find}(x, y) \quad \text{SolutionA} = \begin{pmatrix} 6.455 \\ 10.091 \end{pmatrix} \quad \text{SolutionA} = \begin{pmatrix} 6.45454545 \\ 10.09090909 \end{pmatrix}$$

$$\text{...to 3 decimal places} \qquad \text{...to 8 decimal places}$$

Figure 2

Since the Solve Block routine is an iterative process, changes in the Tolerance setting may affect the time taken to converge to an answer. As well, numerical formatting can be used to output answers accurate to the scale set by the Tolerance. There is no point having a Tolerance of 0.001 and noting an intersection point accurate to 15 decimal places.

Since this process assumes there is a solution to the system and that those solution values of x and y are the same in the first and second equations, the Symbolic Processor's 'Solve for Variable' option can be used effectively after equating the expression for y from the first relation to the expression for *y* from the second. Figure 3 shows the application of the Symbolic Processor. The resulting answer corresponds to the one generated from the Solve Block routine.

$$\frac{90 - 3 \cdot x}{7} = \frac{7 \cdot x - 25}{2} \qquad \text{has solution(s)} \qquad \frac{71}{11}$$

$$\frac{71}{11} = 6.455 \qquad ...\text{and} \qquad y_1\left(\frac{71}{11}\right) = 10.091$$

Figure 3

As the number of dimensions grows, so do the problems of visual presentation. In 3 dimensions, linear relations create flat planes. A system of 3 equations creates 3 planes which, if a finite solution exists, intersect at a point.

$$x + 3y + 2z = 4$$
$$2x + 5y + z = 19$$
$$5x + y + 3z = 12$$

As functions of two variables, *x* and *y,* these can be expressed as (in order of appearance):

$$F(x,y) = \frac{4 - x - 3y}{2} , \; G(x,y) = \frac{19 - 2x - 5y}{1} , \text{ and } H(x,y) = \frac{12 - 5x - y}{3} .$$

These forms are needed as Mathcad's surface plot routine uses a matrix in *x-y* whose elements represent the height *z* of the function above the plane.

Although Mathcad is well suited to representing open 3-dimensional surfaces, it does not seem to allow for the representation of more than one 3-D surface per plot. Figure 4 shows the surfaces created by the first and third equations from the system. The orientation of the axes is equivalent although the scaling differs slightly. A small amount of imagination shows the two planes intersecting in a line.

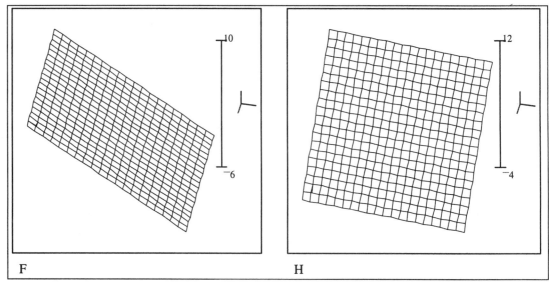

F   H

Figure 4

The point at which the third plane intersects this line is the solution to the system. Obviously, the visual routine breaks down here.

The other tools, 'Solve Blocks' and 'Solve for Variable', still apply. Figure 5 shows the application of the Solve Block utility to the 3x3 system. Again, the Tolerance setting can be adjusted to increase the precision of the output.

$$x := 1 \quad y := 2 \quad z := 3 \quad ...estimates$$

Given

$$x + 3 \cdot y + 2 \cdot z \equiv 4$$
$$2 \cdot x + 5 \cdot y + z \equiv 19$$
$$5 \cdot x + y + 3 \cdot z \equiv 12$$

$$\text{SolutionB} := \text{find}(x, y, z) \qquad \text{SolutionB} = \begin{vmatrix} 4.600 \\ 2.886 \\ -4.629 \end{vmatrix}$$

Figure 5

The use of the 'Solve for Variable' routine is an exercise in typing as the dimensions of the system grow. By equating two of the three equations and by selecting the variable to solve for, an expression is created in the two remaining variables. The process is repeated for another combination of the three original equations. A 2x2 system is generated from these outputs which can then be solved. Although an effective technique, for a high dimension system, the task becomes overwhelming and is equivalent to using the analytical elimination technique. Figure 6 shows the first few steps for the 3-D system already examined.

$$\frac{4 - x - 3 \cdot y}{2} \equiv 17 - 3 \cdot x - 6 \cdot y \text{ has solution(s)} \qquad 6 - \frac{9}{5} \cdot y \qquad \text{from } F(x,y) = H(x,y)$$
with 'x' selected

$$\frac{4 - x - 3 \cdot y}{2} \equiv \frac{12 - 5 \cdot x - y}{3} \text{ has solution(s)} \qquad \frac{12}{7} + y \qquad \text{from } F(x,y) = G(x,y)$$
with 'x' selected

by equating these outputs,

$$\frac{12}{7} + y \equiv 6 - \frac{9}{5} \cdot y \qquad \text{has solution(s)} \qquad \frac{75}{49}$$

a value for 'x' can then be determined from substitution. Likewise for 'z'

Figure 6

At the point where the system dimensions go beyond 3x3, it is not possible to visualize the form of the hyper-surfaces directly. However, just as the solution to pairs of relations reduced the order of a 3x3 system to a 2x2 system, an intermediate solution may be visible as first a surface, then a line, and finally a point.

Explorations

The Basics

1. Solve each of the following systems of linear equations using graphical, numerical and symbolic techiques:

a) $\begin{aligned} y &= -4x + 10 \\ y &= 6x + 5 \end{aligned}$

b) $\begin{aligned} 6R + 5r &= 20 \\ 3R + r &= 5 \end{aligned}$

c)
$$0.4x = 0.8y + 0.5$$
$$0.3y = 0.8 - 0.2x$$

d)
$$6.50x + 3.75y = 5.00$$
$$2.40x - 2.95y = 3.20$$

e)
$$x + 2y - z = 4$$
$$3x - 4y - 2z = -3$$
$$6x - 2y - z = 4$$

f)
$$2x + 3y - 4z + w = 4$$
$$3x - 5y - 5z + 2w = 6$$
$$5x - 2y + 4z - 2w = -1$$
$$-3x + y + z - w = 2$$

2. Verify that each of the following systems of linear equations has no solution.

a)
$$y = 6x + 3$$
$$18x = 3y + 2$$

b)
$$2y + 3x = 15$$
$$y + 5 = -\frac{3}{2}x$$

3. In a simple DC circuit with one 30 V source, the currents can be characterized by the following system:

$$I_1 + I_2 + I_3 = 0$$
$$R_1 I_1 + R_2 I_2 = 30$$
$$-R_2 I_2 + R_3 I_3 = 0$$

where the first relation is the sum of the currents and the next two are sums of voltages. For $R_1 = 1000\ \Omega$, $R_2 = 1000\ \Omega$ and $R_3 = 1500\ \Omega$, determine the solution of this system using a numerical technique. Examine the behavior of the currents as R_3 is made to vary so that $R_3 \gg R_2$.

4. The forces exerted by the two vertical end-supports of a span of length 3.0 m and mass 600 N are defined by $F_1 + F_2 = 600$ N and $F_1 \cdot 1.5 - F_2 \cdot 1.5 = 0$ Nm. Determine the solution of this 2x2 system of forces.

Beyond the Basics

1. The cost analysis of a product reveals a linear relation between quantity and cost of production for the first 500 units. The total relation includes a fixed cost of $5000 for general expenses. Revenues to the 500 unit level are directly related to sales volume. The retail price of the product is $200.00 while the production cost per unit is $125.00. Determine graphically and numerically the point at which the revenue equals the cost of production.

2. Examine the effect an increase or decrease in cost per unit has on the position of the balance point in #1.

3. Create a 3x3 system with coefficients and constants of your choice. Once you have solved the system, change one of the coefficients in an orderly way and examine the effect changes in this one coefficient have on the solution matrix. How would you explain the change or movement in the intersection point in terms of shifting planes? Now choose another coefficient from either the same or another relation. Explain the visual effect of changes to this coefficient.

4. Examine the change in *The Basics* #4 as a 850 N load is placed 1.0 m from the left-hand support. The system force equation changes to $F_1 + F_2 = 1450$ N while the second (torque) relation becomes $F_1 \cdot 1.5 - F_2 \cdot 1.5 + 850 \cdot 0.5 = 0$ N·m for a static span. If the maximum compressive force of each of the two supports is 2000 N determine the weight and position of one possible failure configuration.

4.2 The Use of Determinants

The solution of a 2x2 system can be accomplished quickly and efficiently using the pencil and paper techniques of substitution and elimination. Even though most graphics calculators and mathematics programs (including Mathcad) can allow quick solutions of these systems, the time taken to fire up your system and input the coefficients and constants may not be worth the extra trouble.

However, the situation changes drastically when the system grows to 3x3 and beyond. Anyone exposed to these systems becomes infinitely appreciative of the time saved by using calculation tools. The solution technique then becomes secondary to what the solution is telling you about the behavior of the system.

No doubt, at this stage you have already learned a solution technique called the Method of Determinants or Cramer's Rule as a replacement for substitution and elimination. The application of the method of elimination to a 3x3 system produces a nightmarish combination of constants and indexes which can be coded into the Method of Determinants, a method which reveals the order underlying the seeming confusion of symbols.

Even with this ordered and easily programmed approach, you have probably restricted yourself to 2x2 and 3x3 systems if you have been using pencil and paper. The calculation of the various determinants, while not difficult, requires attention as it is easy to get lost in the details of the generated minor matrices.

The beauty (and advantage) of using Cramer's Rule is that it scales so well. It applies no matter the degree of the system, from 2x2 to 30x30 and beyond.

Other techniques, notably the use of the principal and parallel diagonals of the extended system matrix, hit roadblocks as the size of system grows. The parallel method just does not work. Row reduction, although an interesting exercise into the properties and nature of matrices, can be cumbersome to perform and to program. In this set of explorations, we'll examine the use of determinants within the Mathcad environment.

Warmup

A linear system consists of a group of interconnected relations between the variables. For 3 equations in 3 unknowns, the typical form of each of the three equations, $ax + by + cz = d$, may be more efficiently stated as $\begin{pmatrix} a_1 & b_1 & c_1 \\ a_2 & b_2 & c_2 \\ a_3 & b_3 & c_3 \end{pmatrix} \begin{pmatrix} x_1 \\ x_2 \\ x_3 \end{pmatrix} = \begin{pmatrix} d_1 \\ d_2 \\ d_3 \end{pmatrix}$ where the column of a's represents the coefficients of the x_1 variable, as the columns of b's and c's represent those of the x_2 and x_3 (or x, y and z) variables respectively. The column of d's represents the constants. We will explore the linear algebra of matrix systems in more detail within the next set of exercises (4.3 - Matrix Methods for Systems of Equations).

A matrix in Mathcad is defined by selecting Math/Matrices, or clicking on the Vectors and Matrices Palette icon or using the CTRL-M key combination. The dimensions can then be adjusted to suit the system. Various tools also allow you to choose a particular column or a particular element from the matrix. As well, a submatrix can be extracted and determinants calculated. Matrices can be appended (side-by-side) or stacked (one on top of another).

☑ Mathcad does not seem to allow easy access to row-reduction techniques
without a lot of grief.

Let's examine a Cramer's Rule solution of a 3x3 system as it might be constructed in Mathcad.
First, we must define the matrix of coefficients and the matrix of constants. By manipulating
these matrices and their determinants, we will be able to generate a generic problem solver for
3x3 systems.

And, the method scales easily to systems of higher dimension.

Let the 3x3 system be defined by: $\begin{aligned} x - 2y + 2z &= 5 \\ 5x + 3y + 6z &= 57 \\ x + 2y + 2z &= 21 \end{aligned}$.

Figure 1 shows the definition of the coefficient matrix ABC and the matrix of constants D for the
above system. The ORIGIN variable has been set to one (1). While the default value of this
variable is zero, this change of definition ensures that the first element, column or row of any
matrix is defined as the first (and not the zero'th).

☑ When using indexes, matrices and vectors, define the ORIGIN variable to be
either zero or one depending on your preference. Some of the subroutines used
in Mathcad require a clear and consistent definition of the starting point of an
array.

$$\text{ORIGIN} := 1$$
$$\text{ABC} := \begin{pmatrix} 1 & -2 & 2 \\ 5 & 3 & 6 \\ 1 & 2 & 2 \end{pmatrix} \qquad \text{D} := \begin{pmatrix} 5 \\ 57 \\ 21 \end{pmatrix}$$

Figure 1

By having defined the particular matrix and by using the column selector option (CTRL-6), as in
Figure 2, we may select any column from the matrix (if you need to change a column into a row,
use the CTRL-1 transpose operator). The usefulness of this tool will be seen when we replace a
column of coefficients by the column of constants as defined by Cramer's Rule.

$$\text{ABC}^{<1>} = \begin{pmatrix} 1 \\ 5 \\ 1 \end{pmatrix} \qquad \text{ABC}^{<2>} = \begin{pmatrix} -2 \\ 3 \\ 2 \end{pmatrix} \qquad \text{ABC}^{<3>} = \begin{pmatrix} 2 \\ 6 \\ 2 \end{pmatrix}$$

Figure 2

As well we may place matrices or parts of matrices beside one another using the AUGMENT
function as long as the number of rows matches. The STACK command places matrices atop one
another if there is a column number match. Figure 3 shows both of these commands in action
using the previously defined square matrix ABC.

$$\text{augment}(ABC, ABC) = \begin{pmatrix} 1 & -2 & 2 & 1 & -2 & 2 \\ 5 & 3 & 6 & 5 & 3 & 6 \\ 1 & 2 & 2 & 1 & 2 & 2 \end{pmatrix} \qquad \text{stack}(ABC, ABC) = \begin{bmatrix} 1 & -2 & 2 \\ 5 & 3 & 6 \\ 1 & 2 & 2 \\ 1 & -2 & 2 \\ 5 & 3 & 6 \\ 1 & 2 & 2 \end{bmatrix}$$

Figure 3

Unfortunately, the AUGMENT and STACK functions do not allow more than 2 arguments within their parentheses. This is an obstacle. And, as always, obstacles make for interesting discoveries. How would you combine 3 reordered columns taken from the matrix ABC?

Luckily, the AUGMENT function does allow itself to be embedded within other AUGMENT functions. Figure 4 shows the replacement of the first column within the matrix of coefficients with the column of constants using the embedding technique (a). A second utility, SUBMATRIX is applied with the same result (b). However, while the first works for any column replacement, the second method works most easily on the outer columns only.

The syntax of the SUBMATRIX command is:
submatrix(name, start row, finish row, start column, finish column)

embedded augment functions

$$\text{augment}\left(\text{augment}\left(D, ABC^{<2>}\right), ABC^{<3>}\right) = \begin{pmatrix} 5 & -2 & 2 \\ 57 & 3 & 6 \\ 21 & 2 & 2 \end{pmatrix}$$

extraction of square submatrix from ABC

$$\text{submatrix}(ABC, 2, 3, 2, 3) = \begin{pmatrix} 3 & 6 \\ 2 & 2 \end{pmatrix}$$

use of submatrix in column replacement

$$\text{augment}\left(D^{<1>}, \text{submatrix}(ABC, 1, 3, 2, 3)\right) = \begin{pmatrix} 5 & -2 & 2 \\ 57 & 3 & 6 \\ 21 & 2 & 2 \end{pmatrix}$$

Figure 4

As yet we still have not solved the system. However, with the tools at hand, we are ready to calculate the determinants of the required matrices and form their ratios. Figure 5 shows the application of the determinant command to the matrix ABC. The determinant is invoked by typing " | " followed by the name of the matrix.

Once the determinants have been defined, they can be used to form quotients consistent with the solutions for x, y and z (or x_1, x_2 and x_3).

$$\text{del} := \big|ABC\big|$$

$$\text{delx} := \Big|\text{augment}\Big(\text{augment}\big(D, ABC^{<2>}\big), ABC^{<3>}\Big)\Big|$$

$$\text{dely} := \Big|\text{augment}\Big(\text{augment}\big(ABC^{<1>}, D\big), ABC^{<3>}\Big)\Big|$$

$$\text{delz} := \Big|\text{augment}\Big(\text{augment}\big(ABC^{<1>}, ABC^{<2>}\big), D\Big)\Big|$$

Figure 5

And, now that the determinants have all been properly defined, the solutions can be evaluated as shown in Figure 6.

$$\text{del} = 16 \qquad \text{delx} = 48 \qquad \text{dely} = 64 \qquad \text{delz} = 80$$

$$\text{solx} := \frac{\text{delx}}{\text{del}} \quad \text{solx} = 3 \qquad \text{soly} := \frac{\text{dely}}{\text{del}} \quad \text{soly} = 4 \qquad \text{solz} := \frac{\text{delz}}{\text{del}} \quad \text{solz} = 5$$

Figure 6

You may well ask, after all this hard work and manipulation, whether this is easier than solving the system by the old reliable way? Of course, if you have only one system to solve, the race may end in a tie. The strength of creating a generic 3x3 system solving tool is in its application to the next problem. And the next. Your half-page program begins to shine. By quickly editing the matrices to suit the coefficient and constants of the new system, the solution emerges at the push of the F9 button.

An even quicker design (to edit repeatedly) may include a menu area at the top of the file into which the new constants and coefficients may be placed. Each of these can be scaled to suit any dimension system.

Explorations

The Basics

1. Use the determinant method to solve each of the following 2x2 linear systems.
   a) $\begin{aligned} 0.65x_1 + 0.88x_2 &= 7.0 \\ 0.24x_1 - 0.56x_2 &= 2.5 \end{aligned}$
   b) $\begin{aligned} 7x + 4y &= 10 \\ 2y &= 3x + 1 \end{aligned}$

2. Use the determinant method to solve each of the following 3x3 linear systems
   a) $\begin{aligned} 2x + 3y - 6z &= 3 \\ 3x - 4y - z &= 2 \\ 9x - 6y + 5z &= 40 \end{aligned}$
   b) $\begin{aligned} 2x - 2y + 2z &= 4 \\ x + y - z &= 100 \\ 3x + y - 2z &= 6 \end{aligned}$

3. Create a generic 3x3 linear system solver using the method outlined in the Warmup. To save typing repeated statements, make use of the 'cut, copy and paste' (CTRL-X, CTRL-C, CTRL-V) operators common to all Windows programs.

4. Create a generic 3x3 linear system solver using the top portion of the page as a definition area for the various coefficients and constants (i.e. $a_1 := \qquad$, $a_2 := \qquad$, $a_3 := \qquad$). Does this style offer any advantages over that from *The Basics* #1?

Beyond the Basics

1. Use a generic 5x5 system solver to solve the following analysis of an electrical circuit containing multiple current loops. *I* represents the current (in Amperes) in each loop while the coefficients of *I* represent resistances (in Ohms) and the constants, on the right-hand side of the second, fourth and fifth equations, represent voltages (in Volts).

$$I_1 + I_2 + I_3 + 0 + 0 = 0$$
$$0 + 200I_2 + 0 + 0 + 0 = 60$$
$$0 + 0 + I_3 + I_4 + I_5 = 0$$
$$0 - 200I_2 + 1500I_3 + 0 + 200I_5 = 0$$
$$0 + 0 + 0 + 500I_4 - 200I_5 = 0$$

2. Determine the new solution of the 5x5 system in *Beyond the Basics* #1 if the I_3 resistance from the fourth equation is changed from $+1500\Omega$ to $250\ \Omega$.

3. Edit one of *The Basics* #1 or #2 to create a generic 4x4 system solver. Test its effectiveness against the Solve Block routines created in Section 4.1

4.3 Matrix Methods for Systems of Equations

The numerical methods we have explored so far have taken advantage of graphical estimation and the speed of numerical iterations. The first technique allowed an overall picture of the system while the quick looping of the second routine generated solutions to the desired accuracy.

A more analytic approach examined the creation of solutions using Cramer's Rule, an organized variant of the elimination method.

Still, we have not taken advantage of the basic linearity of the equations within the system. Let's return to the initial generation of the matrix format from the more familiar linear form.

For a 2x2 system, the general form of the equations is:

$$a_{11}x_1 + a_{12}x_2 = b_1$$
$$a_{21}x_1 + a_{22}x_2 = b_2$$

where the indexed a's represent the coefficients of the variables and the indexed b's represent the constants.

> ☑ The choice of an indexed x (as in x_1, x_2 ,...) variable over the use of x and y (and z and so on) variable names generates a more consistent design easily scaled up to higher order systems.

The system may be more compactly stated in a matrix form as $\begin{pmatrix} a_{11} & a_{12} \\ a_{21} & a_{22} \end{pmatrix} \begin{pmatrix} x_1 \\ x_2 \end{pmatrix} = \begin{pmatrix} b_1 \\ b_2 \end{pmatrix}$ where the indexes follow a row-column format.

The two forms are equivalent. One of the advantages of the matrix form is the clear way in which coefficients are differentiated from variables and constants. Each quantity has its separate well-defined area.

As a linear system, the matrices may then be represented by $Ax = B$, where each of A, B and x represent a separate matrix. This form parallels the more familiar structure of a single linear equation, itself simply a 1x1 system.

For a linear equation, the solution of $Ax = B$ would be $x = \dfrac{A}{B}$. This simple process of division by A needs careful attention when applied to matrices. The matrix process treats division as multiplication by the reciprocal. The challenge of finding the reciprocal matrix is a worthy one and, if done by hand, gives you a profound respect for determinants and minor matrices.

However, systems with dimensions above 2x2 are overpoweringly tedious. Luckily, the advent of computer symbolic and numerical systems has pretty much done away with the raw techniques involved and has allowed more time for playing with the system elements themselves.

Warmup

The matrix operations applied to linear systems parallel those applied to a single linear equation with a few important exceptions.

The addition of two real numbers is a commutative process. The dimension (one) of each number matches that of the other. The addition of matrices is also a commutative process as long as the dimensions of the matrices are equal.

A matrix is created in Mathcad by selecting Math/Matrix, clicking on the Vectors and Matrices Palette icon or by typing the key combination CTRL-M. The dimensions of the matrix can be chosen and the empty place holders filled within the array by either using the mouse or the TAB key to move between place holders.

The matrix can exist on its own within a calculation or can appear on the right-hand side of a definition statement. In Figure 1, a 3x3 matrix has been defined as A while a 3x1 matrix has been defined as B. An attempt to add the two results in an error message. Since addition is a like-index element to element process, the first combination returns an error message. There are simply not enough elements defined in the second matrix. Or, there are too many in the first matrix. Either way, there is a mismatch.

$$A := \begin{pmatrix} 2 & 6 & 6 \\ 4 & -7 & 7 \\ 5 & 5 & 8 \end{pmatrix} \qquad B := \begin{pmatrix} 3 \\ 4 \\ 5 \end{pmatrix} \qquad A+B= \boxed{\text{array size mismatch}}$$

Figure 1

For the matrices in Figure 1, reverse the order of addition and see if the problem clears up.

There has to be a dimensional match for addition (or subtraction) to take place. This makes sense as the elements A_{12} or A_{33} search for like index elements in the second array. Since there are no equivalent B_{12} or B_{33} elements, respectively, addition is undefined.

Don't confuse the non-existence of an element with an element equaling zero. Although there may be an argument for filling up the holes with zeros, the two systems given in Figure 1 represent dimensionally different quantities.

For a matched pair of matrices, as shown in Figure 2, the addition process takes place without problem. Check for yourself that like elements have been added and subtracted and that the process is commutative.

$$A := \begin{pmatrix} 2 & 6 & 6 \\ 4 & -7 & 7 \\ 5 & 5 & 8 \end{pmatrix} \quad D := \begin{pmatrix} 4 & 4 & 2 \\ 7 & 5 & 8 \\ 8 & 3 & 7 \end{pmatrix} \quad A+D = \begin{pmatrix} 6 & 10 & 8 \\ 11 & -2 & 15 \\ 13 & 8 & 15 \end{pmatrix} \quad A-D = \begin{pmatrix} -2 & 2 & 4 \\ -3 & -12 & -1 \\ -3 & 2 & 1 \end{pmatrix}$$

Figure 2

The scalar multiplication of a matrix distributes the scalar to each element of the matrix as it should since it scales the matrix up or down depending on its size and sign. Figure 3 shows the result of scalar multiplications and a subsequent addition applied to matrices A and D.

$$4 \cdot A = \begin{pmatrix} 8 & 24 & 24 \\ 16 & -28 & 28 \\ 20 & 20 & 32 \end{pmatrix} \quad 3 \cdot D = \begin{pmatrix} 12 & 12 & 6 \\ 21 & 15 & 24 \\ 24 & 9 & 21 \end{pmatrix} \quad 4 \cdot A - 3 \cdot D = \begin{pmatrix} -4 & 12 & 18 \\ -5 & -43 & 4 \\ -4 & 11 & 11 \end{pmatrix}$$

Figure 3

The multiplication of two matrices (the process we are more interested in for system solutions) requires careful attention to the matching of the dimensions between the two (or more) matrices. The process of ordinary multiplication of two numbers is commutative ($a \cdot b = b \cdot a$). The same does not apply for matrices.

The multiplication of two matrices A and B of dimension 2x2 and 2x1 respectively would be given by:

$$\begin{pmatrix} a_{11} & a_{12} \\ a_{21} & a_{22} \end{pmatrix} \begin{pmatrix} b_1 \\ b_2 \end{pmatrix} = a_{11} \cdot b_1 + a_{12} \cdot b_2 + a_{21} \cdot b_1 + a_{22} \cdot b_2 .$$

Note the match of the number of columns of the first matrix with the number of rows of the second. If the order of the matrices is reversed, the match is broken and multiplication cannot proceed. The process of multiplying a 3x3 matrix by a 3x1 matrix outputs a 1x3 matrix whose elements are the sums of the products of the row elements and corresponding column elements.

Figure 4 shows both this process and an attempt to reverse the order of the matrices.

$$A := \begin{vmatrix} 2 & 6 & 6 \\ 4 & -7 & 7 \\ 5 & 5 & 8 \end{vmatrix} \quad B := \begin{vmatrix} 3 \\ 4 \\ 5 \end{vmatrix} \quad A \cdot B = \begin{vmatrix} 60 \\ 19 \\ 75 \end{vmatrix} \quad \begin{array}{c} B \cdot A \\ \boxed{\text{array size mismatch}} \end{array}$$

$$C := (3 \ 4 \ 5) \quad C \cdot A = (47 \ 15 \ 86) \quad \begin{array}{c} A \cdot C \\ \boxed{\text{array size mismatch}} \end{array}$$

Figure 4

The multiplication of (*A* x *B*) results in a match of 3x3 by 3x1 and results in a 3x1 matrix. An attempt to multiply a 3x1 matrix by a 3x3 matrix (*B* x *A*) results in an error message. The reorganization of the elements of *B* into a single row matrix *C* allows (*C* x *A*) but not (*A* x *C*) for the same reason of mismatching.

Figure 5 shows the 3x3 matrix *A* multiplying a like-dimension matrix *D*. Note that although multiplication proceeds for both (*A* x *D*) and (*D* x *A*), matrix multiplication is not commutative.

$$A := \begin{vmatrix} 2 & 6 & 6 \\ 4 & -7 & 7 \\ 5 & 5 & 8 \end{vmatrix} \quad D := \begin{vmatrix} 4 & 4 & 2 \\ 7 & 5 & 8 \\ 8 & 3 & 7 \end{vmatrix} \quad A \cdot D = \begin{vmatrix} 98 & 56 & 94 \\ 23 & 2 & 1 \\ 119 & 69 & 106 \end{vmatrix} \quad D \cdot A = \begin{vmatrix} 34 & 6 & 68 \\ 74 & 47 & 141 \\ 63 & 62 & 125 \end{vmatrix}$$

Figure 5

What about division? So far, matrix behavior has paralleled the algebra of a single linear equation fairly closely. Does the same work for matrix division? Figure 6 shows an attempt to divide *E* (the product of *A* and *D*) by *D* directly. The process is undefined.

$$E := A \cdot D \quad E = \begin{vmatrix} 98 & 56 & 94 \\ 23 & 2 & 1 \\ 119 & 69 & 106 \end{vmatrix} \quad \begin{array}{c} \dfrac{E}{D} = \\ \boxed{\text{illegal array operation}} \end{array}$$

Figure 6

This may seem a strange result. After all, the multiplication of two 3x3 matrices as given in Figure 4 yielded a third 3x3 matrix. The division of this resulting matrix E by D should return the matrix A. If $A \cdot D = E$ then $A = \dfrac{E}{D}$. Yet, this is not the case with matrices. However, if you return to your first encounters with linear algebra, you will see that the following process is consistent with the definition of division.

The process of division in real numbers is identical to the process of multiplying by a reciprocal. And, a number times its reciprocal gives one or the identity (with the obvious exception of zero).

In matrix algebra, the reciprocal of the matrix must be defined if division is to be replaced by an equivalent multiplication process.

The matrix A is said to have an inverse A^{-1} if a matrix exists such that $A \cdot A^{-1} = A^{-1} \cdot A = I$ where I is the unit or identity matrix. Figure 7 shows the result of determining the inverse of the matrix D and includes a check of its accuracy.

The unit matrix, defined by having diagonal elements of one and off-diagonal elements of zero, multiplies any other matrix of matched dimension and simply returns the other matrix unchanged. You can check this for yourself.

$$D^{-1} = \begin{vmatrix} 0.167 & -0.333 & 0.333 \\ 0.227 & 0.182 & -0.273 \\ -0.288 & 0.303 & -0.121 \end{vmatrix} \qquad D^{-1} \cdot D = \begin{vmatrix} 1 & 0 & 0 \\ 0 & 1 & 0 \\ 0 & 0 & 1 \end{vmatrix} \qquad D \cdot D^{-1} = \begin{vmatrix} 1 & 0 & 0 \\ 0 & 1 & 0 \\ 0 & 0 & 1 \end{vmatrix}$$

Figure 7

Now that we have defined the inverse matrix of D, we can apply it to the $(A \times D)$ product output E and see whether of not A is generated. Since multiplication is not commutative, we must take care with the order in which the inverse is applied. For the linear relation $A \times D = E$, dividing each side of the equation by D is equivalent to multiplying each side of the equation from the right by D^{-1}. Then we have,

$$\text{original relation } A \cdot D = E$$
$$\text{multiply by inverse } A \cdot D \cdot D^{-1} = E \cdot D^{-1}$$
$$\text{reduce to identity } A \cdot I = E \cdot D^{-1}$$
$$\text{emergence of } A = E \cdot D^{-1}$$

Figure 8 confirms that this works for the matrices A, D and E (the product of A and D).

$$E \cdot D^{-1} = \begin{vmatrix} 2 & 6 & 6 \\ 4 & -7 & 7 \\ 5 & 5 & 8 \end{vmatrix} \qquad \textit{versus} \qquad A = \begin{vmatrix} 2 & 6 & 6 \\ 4 & -7 & 7 \\ 5 & 5 & 8 \end{vmatrix}$$

Figure 8

All of which finally brings us to systems and their solutions. If we are able to express a system of N equations in N unknowns in terms of a NxN matrix of coefficients A, a Nx1 matrix of variables x and a Nx1 matrix of constants B such that $Ax = B$, then, assuming an inverse of A exists and is applied from the left to both sides of the equation, $A^{-1} A x = A^{-1} B$.

Since $A^{-1} A = I$ then $x = A^{-1} B$.

The solution matrix is generated from the inverse of the coefficient matrix times the constant matrix. Figure 9 shows this solution process for a simple 3x3 system.

$$A := \begin{pmatrix} 3 & 6 & 7 \\ 4 & 6 & 9 \\ 5 & 8 & 7 \end{pmatrix} \quad \text{matrix of coefficients} \qquad B := \begin{pmatrix} 10 \\ 12 \\ 14 \end{pmatrix} \quad \text{matrix of constants}$$

inverse matrix and check

$$A^{-1} = \begin{pmatrix} -1.154 & 0.538 & 0.462 \\ 0.654 & -0.538 & 0.038 \\ 0.077 & 0.231 & -0.231 \end{pmatrix} \qquad A \cdot A^{-1} = \begin{pmatrix} 1 & 0 & 0 \\ 0 & 1 & 0 \\ 0 & 0 & 1 \end{pmatrix}$$

$$x := A^{-1} \cdot B \qquad \text{definition of solution} \qquad x = \begin{pmatrix} 1.385 \\ 0.615 \\ 0.308 \end{pmatrix} \quad \text{solution to system}$$

check of solution within system

$$A \cdot x - B = \begin{bmatrix} 1.776 \cdot 10^{-15} \\ -3.553 \cdot 10^{-15} \\ 0 \end{bmatrix}$$

Figure 9

Note that the match is not perfect but generates an error on the order of 10^{-15}. Is there a way of reducing this to zero? Well, you can always redefine the zero to be 10^{-14} ! Remember, these are numerical routines and prone to inaccuracies.

Explorations

The Basics

1. Perform the following operations on the matrices A and B:

$$A = \begin{pmatrix} -4 & 34 & -9 \\ 64 & 8 & 34 \\ 5 & 76 & 12 \end{pmatrix} \text{ and } B = \begin{pmatrix} -3 & -4 & 9 \\ 17 & -54 & 35 \\ 17 & 0 & -23 \end{pmatrix}$$

a) $A + 3B$ b) $A\,B$ c) $B\,A$ d) B^2

e) $B^{-1} A\,B$ f) $I\,A$ g) $B^{-1} B\,A$

2. Show that the quadratic expression $(A^2 - B^2\,I)$, where A is a square matrix, B is a scalar constant and I is the unit matrix, factors to $(A - BI)(A + BI)$. Does this work for any NxN matrix A?

3. Solve the following systems of equations using matrix algebra:

a)
$$3.5x - 2.4y = 4.5$$
$$6.5y + 2.9x = 3.4$$

b)
$$3I_1 + 10I_2 = 0$$
$$6I_1 - 4I_2 = 10$$

c)
$$x_1 + 3x_2 + 4x_3 = 0$$
$$2x_1 - 4x_2 + x_3 = -4$$
$$-x_1 + 4x_2 + 4x_3 = 15$$

4. For the following 3x3 system, $\begin{pmatrix} 23 & 7 & -6 \\ -9 & 0 & 8 \\ 8 & -65 & 3 \end{pmatrix} \cdot x = \begin{pmatrix} -2 \\ 7 \\ -2 \end{pmatrix}$ check whether the matrix

$x = \begin{pmatrix} 0.16 \\ 0.10 \\ 1.05 \end{pmatrix}$ is a possible solution. Determine the numerical solution (to 15 decimal

place accuracy) and compare.

5. Solve the following 5x5 system using matrix algebra.

$$\begin{pmatrix} 3 & 8 & 9 & 4 & 9 \\ 2 & -7 & 45 & 4 & 6 \\ -9 & -9 & 4 & 4 & -5 \\ 4 & 5 & 65 & 4 & -8 \\ 65 & 20 & 2 & 4 & -4 \end{pmatrix} \cdot x = \begin{pmatrix} 0 \\ 4 \\ -6 \\ 0 \\ 5 \end{pmatrix}$$

Remember to check for the validity of the inverse. Does the change of one of the coefficients by 5% create like changes in the solution?

Beyond the Basics

1. Give 3 examples of linear matrix systems without solutions.
2. Does the TOL (tolerance) setting or the numerical precision affect the accuracy of the solution to a system or the time taken in its calculation? Give reasons for your answer.
3. Appendix 1 examines the rotation of a coordinate system about the origin. A point (x_0, y_0) in the coordinate system x-y has the coordinates (x_0', y_0') in a new system x'-y' whose axes are rotated a counter-clockwise angle θ about the origin. The

   transformation is accomplished using the matrix relation $\begin{pmatrix} x_0' \\ y_0' \end{pmatrix} = \begin{pmatrix} \cos\theta & \sin\theta \\ -\sin\theta & \cos\theta \end{pmatrix} \begin{pmatrix} x_0 \\ y_0 \end{pmatrix}$.

   For the point (-6,+5) in the x'-y' reference frame, determine the original coordinates if the x'-y' system's rotation is +125° with respect to the x-y reference frame. What is the general form of the inverse process, the evaluation of the original coordinates from the rotated coordinates? Evaluate the determinant of the transformation matrix.
4. For the 2x2 matrix system below, develop a method of repeated but directed trial and error which would converge to the solution after a fixed number of iterations.

   $\begin{pmatrix} 3.8 & 7.3 \\ 2.2 & -5.2 \end{pmatrix} \begin{pmatrix} x_1 \\ x_2 \end{pmatrix} = \begin{pmatrix} 3.0 \\ 3.2 \end{pmatrix}$ with a starting point of $\begin{pmatrix} 1.000 \\ -0.200 \end{pmatrix}$.

Chapter 5: Statistics

There is not enough time in the universe to measure a characteristic of all members of a population whether it be a population of fruit flies, of people, the currents and voltages across a resistor in a laboratory or the dimensions of all industrial products coming off an assembly line.

The process of taking a sample from the larger system and using this information as a representation of the system's behavior suggests a lack of completeness in our information. For the sampling process, we infer behavior rather than describe it. However, if we have analyzed our data completely and fairly (without bias), the newly generated information or relationships should allow for rational decisions and intelligent predictions. They display a high level of confidence.

Mathcad offers a number of tools (utilities and operators) which can severely lessen the burden of performing statistical analysis. The problem has always been in the masses of data you wish to analyze but do not necessarily want to process by hand. In fact, no one *does* statistics by hand any more. Once the theory is understood, its repeated application to 5,000 or 10,000 data points would filter all joy and interest from the topic. Better to let a well-designed tool handle the repetitive labor and concentrate on the analysis of the data and the meaning of the results.

5.1 Measures of Central Tendency and Variance

Data is the starting point. A parameter or indicator is the end point. As the raw data collected goes through the process of analysis, a representative number appears. If you have polled all members of the system under inspection, then the generated number describes a characteristic of that population (descriptive statistics). If, however, you have only polled a smaller portion or sample of the larger population, the best your final result can do is infer a characteristic (inductive statistics).

In this set of explorations, we will concentrate on some of the tools of statistical analysis Mathcad offers. These will allow the transformation of raw data into a representative number. In practice, the raw data should come from measurements within a population (either complete or sampled) or by the repeated measurement of the same variable within a system (as in the case of physical error measurement).

The data can be input into the analyzer (here, Mathcad) point by point by hand, can be generated from a functional relationship (which defeats the purpose since you know the answer before having read the question) or be read into the analysis environment from an unstructured or structured data file. Within this set of explorations, we will generate the data from prescriptions. Realize, however, that this presents a simulation of the process and that the data should be generated from experience and measurement.

Warmup

READ and WRITE Operators

If our data represented the heights of people as the only variable within a population then the information can simply be stored as one measurement after another. We can use Mathcad to write this information into a data file for later retrieval and analysis. The measurements stored in the text file are simply separated by a space. If your data is more structured, organized into a matrix, then the tools described in Section 9.4 (READPRN, WRITEPRN) are required.

To analyze the data, the information must be read into the Mathcad environment either as a string of numbers or as a matrix. In the process you need information regarding the number of data points within the file. This can be read from a simple text file or can be generated using the **length**(vector_name) operator. Figure 1 shows the process of writing the data to a file while Figure 2 shows the inverse process of reading. Note the number of data points has been determined two different ways.

In Figure 1, the data file is written to Mathcad's default directory. The data can be checked by examining the file in any text reader/editor.

Note the vector index begins at 0. There are $N+1$ data points. Any confusion (# of data points versus index #) can be reduced by setting the ORIGIN variable to 1 at the top of your worksheet so as to redefine the first element of the vector as index number 1.

Some care must be taken to limit the number of characters in the filename to no more than 8 (along the DOS 8.3 filename convention). You *can* have variable names greater than 8 characters. However, the data file itself will have only a truncated version of this full name. In our case, the height data file would have been saved as height_d.dat in the Mathcad directory. The same applies for the READ cycle. Only the first 8 characters are compared.

If you have edited the worksheet and wish to update or test the **READ/WRITE** process, select Math/Calculate Worksheet from the menu.

$N := 6$ number of (data points/measured people) - 1

$k := 0.. N$ vector index

$WRITE(hitenumb) := N$ write number to file (optional)

$height_k :=$

data $WRITE(height\_data) := height$ write height data to file

1.25
1.76
1.55
1.68
1.95
2.05
2.20

Figure 1

$number := READ(hitenumb)$ read the number data from file (optional)

$number = 6$ $j := 0.. number$ (number of records - 1) and index

$read\_height_j := READ(height\_data)$ read height data from file

$numberA := length (read\_height)$ number of records using length(v)

$numberA = 7$ operator instead of text file

$read\_height_j$

1.250
1.760
1.550
1.680
1.950
2.050
2.200

Figure 2

Simple Statistical Operators

In Figure 3, the operators **min**(x), **max**(x), **sort**(x), **mean**(x) have been applied to the data, where:

- **min**(x) outputs the minimum value from the vector x
- **max**(x) outputs the maximum value from the vector x
- **sort**(x) outputs the vector in ascending order
- **mean**(x) outputs the arithmetic mean of the group of elements or vector

The arithmetic mean has also been calculated using its explicit definition as shown at the bottom of Figure 3.

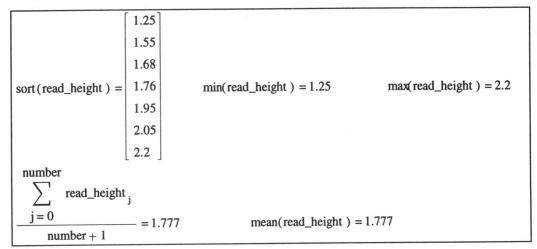

$$\text{sort}(\text{read\_height}) = \begin{bmatrix} 1.25 \\ 1.55 \\ 1.68 \\ 1.76 \\ 1.95 \\ 2.05 \\ 2.2 \end{bmatrix} \qquad \min(\text{read\_height}) = 1.25 \qquad \max(\text{read\_height}) = 2.2$$

$$\frac{\displaystyle\sum_{j=0}^{\text{number}} \text{read\_height}_j}{\text{number}+1} = 1.777 \qquad \text{mean}(\text{read\_height}) = 1.777$$

Figure 3

☑ The Σ operator is either selected from the icon menu (Version 5.0), from the Calculus palette (Version 6.0) or by typing $ in either version. The read_height vector can also be summed using CTRL-4 (read_height).

Another measure of the central tendency of a distribution of data points is the median, or middle data point for the sorted array. We have already sorted our data in ascending order. And, since there are 7 points, the median must be the 4th element (i.e., element read_height$_3$ with an ORIGIN of 0) of the sorted array. Figure 4 confirms this.

$$\text{sort}(\text{read\_height}) = \begin{bmatrix} 1.25 \\ 1.55 \\ 1.68 \\ 1.76 \\ 1.95 \\ 2.05 \\ 2.2 \end{bmatrix} \qquad \text{median}(\text{read\_height}) = 1.76$$

Figure 4

However, what happens if the number of data points is even. Strictly speaking, there is no middle point then. Then the average of the two points adjacent to the mid point is taken as illustrated in Figure 5. Another data point (1.73) has been inserted between the data points 1.68 and 1.76. The min and max outputs have not changed while the mean has shifted slightly. However, the median point is no longer taken directly from the data but rather is a calculated value.

$$\text{mean}(\text{height}) = 1.771 \qquad \min(\text{height}) = 1.25 \qquad \max(\text{height}) = 2.2$$

$$\text{median}(\text{height}) = 1.745 \qquad \frac{1.73 + 1.76}{2} = 1.745$$

Figure 5

By now you are asking yourself, "So what?" Everything we have done so far could easily have been done with a piece of paper and a dull pencil in half the time. What if your data includes 50 points? 500 points? 5000 points? Then, the operators Mathcad offers may indeed seem a little more worthy of attention.

So far we haven't visually represented our data. And there really hasn't been any need to. The numbers, separate as they are, stand well on their own.

Frequency Distributions

In larger distributions the same measured numbers may appear with a frequency greater than 1. In the case of our present example, the measurements of all people within a city population would probably result in a large number of people at each height measurement. Instead of keeping their separateness, they may be grouped into classes or intervals, a range of values into which a measurement within adds one to the count. Frequency distributions then represent a two dimensional data, with the range of values along the horizontal axis and the number of times an element of the population falls within that range as the output along the vertical axis.

In Figure 6, random numbers have been generated for each point along the horizontal axis. The output of the raw data is shown by x-marks. Minimum and maximum values have been generated as well. The arithmetic mean offered by Mathcad adds up all the data and divides by the number of data points. As we transform this data into a frequency distribution, the definition of the mean changes slightly. Although it becomes less accurate, the computation involved is reduced.

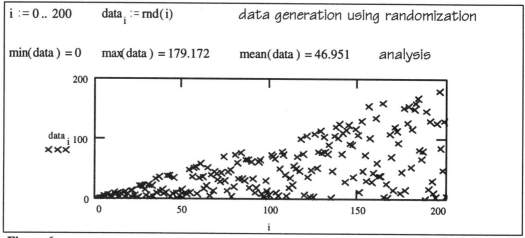

Figure 6

As can be seen from the scatter plot above (which may be different from the one you create due to the **rnd**(x) operator), at each point along the *x*-axis a corresponding random value has been generated. If we break the *y*-axis into a set of intervals, then the number of data points having output values within each interval can be measured and a frequency distribution generated. For each interval, the frequency represents the number of times a number falls within that range. This process is equivalent to looking at Figure 6 on its side and counting the number of x's within each interval. Figure 7 does just that. The number of x's between the vertical lines defining the intervals represent the frequency of occurence for numbers within a particular interval.

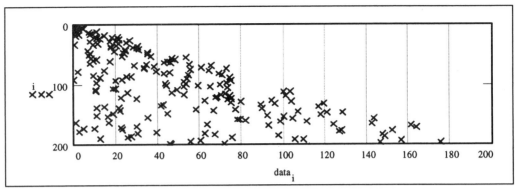

Figure 7

As in Figure 7, an interval of width 20 was chosen in Figure 8. There are 10 bins into which the data can fit. Mathcad offers the **hist**(interval, data) operator which uses the vector interval information to process the vector data. The plots represent the same information presented as
    a) a frequency polygon and
    b) as a histogram.

Since the intervals are a length of 20, the frequency polygon points and the histogram bars are centered over the midpoint of the interval (i.e., at 10, 30, 50, ...) along the horizontal axis.

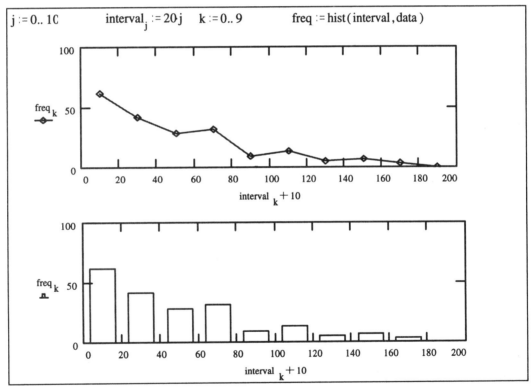

Figure 8

Measures of Variance

By representing the behavior of a system by a single number (or a set of numbers) information is both gained and lost. The individual data points fall into secondary place as the representative measure takes over. In the case of the height measurements, although the mean and median do indicate the overall central tendency of heights within our population (or sample), there may have been extreme instances. Two sets of data which have the same mean (and median) may be representative of very different populations. A simple example is given in Figure 9.

$$k := 0..3 \qquad A_k := \qquad\qquad B_k :=$$

A_k	B_k
19	0
20	40
24	4
25	44

$$\text{mean}(A) = 22 \qquad\qquad \text{mean}(B) = 22$$
$$\text{median}(A) = 22 \qquad\qquad \text{median}(B) = 22$$

Figure 9

In some way, the lost information must be regained. Some knowledge of the overall distribution of the points that were used in the calculation of the mean must be attached to the mean for it to make any sense. This is the purpose of the variance and its square root, the standard deviation. They offer a measure of the spread of the raw data from the mean.

Mathcad includes the operators **var**(x) and **stdev**(x) which operate on unstructured data (the vector x) and produce measures of the overall scatter of points about the mean. However, care must be taken when applying these ideas to the frequency distributions. In this case, the functions can be defined so that no confusion results.

If we apply these operators to the original 7 points of height data, we generate a number which indicates the relative distance of the points from the mean. The standard deviation is also defined explicitly in Figure 10 as:

> the squareroot of the difference of the average of the square and the square of the average.

A more digestible form is: $s = \sqrt{\dfrac{\sum x^2}{N} - \left(\dfrac{\sum x}{N}\right)^2}$ where N represents the number of points.

$$\text{number} + 1 = 7 \qquad\qquad \text{\# of data points}$$
$$\text{mean}(\text{read\_height}) = 1.777 \qquad \text{average of heights}$$
variance and standard deviation
$$\text{var}(\text{read\_height}) = 0.089 \qquad\qquad \text{Mathcad operators}$$
$$\text{stdev}(\text{read\_height}) = 0.298$$

$$\text{standarddev} := \sqrt{\frac{\displaystyle\sum_{j=0}^{\text{number}} \left(\text{read\_height}_j\right)^2}{\text{number} + 1} - \left[\frac{\displaystyle\sum_{j=0}^{\text{number}} \text{read\_height}_j}{\text{number} + 1}\right]^2}$$

$$\text{standarddev} = 0.298$$

Figure 10

The standard deviation of 0.298 indicates most of the data points fall within the range (1.777 - 0.298) to (1.777 + 0.298). If the population were to grow, we can be fairly certain that no one will appear with a height of 4.0 m or 0.2 m. These are outside the pattern of the original data.

The data in Figure 10 is equally weighed. No one data point has any more importance than another. However, in a frequency distribution, the interval midpoints each have a distinct weight defined by the frequency. The definition of mean must then change to take this into account. Applying the Mathcad operator **mean**(x) to the frequency distribution will not produce a result consistent with the distribution.

For our frequency distribution data (Figures 6, 7 and 8), the mean is defined as $\dfrac{\sum f \cdot x}{\sum f}$ with x representing the interval midpoint and f, its frequency. The weighted average produces a result consistent with the frequency representation. Figure 11 illustrates this definition.

$$\text{freqmean} := \frac{\sum\limits_{k} \text{plot}_k \cdot \left(\text{interval}_k + 10\right)}{\sum\limits_{k} \text{plot}_k} \qquad\qquad \text{freqmean} = 49.602$$

Figure 11

The exact arithmetic mean for this collection of points is 46.951 from Figure 6. As a frequency distribution, the same data displays a central tendency near 50. While the first method gave equal weight to each point, the second grouped them into classes with a number, the frequency, measuring their rate of occurence. The means are close enough to at least show the consistency of the method.

Even though data points range between 0 and 200, the lower value points carry more weight. There are more of them and they drag the mean below the mid-point of the range at 100.

In the same manner, the standard deviation must be applied so that the frequencies are taken into account. The square of the average and the average of the square must be frequency dependent.

The standard deviation s is then given by: $\quad s = \sqrt{\dfrac{\sum fx^2}{\sum f} - \left(\dfrac{\sum fx}{\sum f}\right)^2}\quad$ or its equivalent as defined

in Figure 12. Here, the definitions of the variance and standard deviation in general have been applied to the random number distribution we have been examining.

$$\text{midpoint}_k := \text{interval}_k + 10 \qquad \text{redefined midpoint}$$

$$\text{freqvar} := \frac{\sum\limits_{k} \text{freq}_k \cdot \left(\text{midpoint}_k - \text{freqmean}\right)^2}{\sum\limits_{k} \text{freq}_k} \qquad\qquad \text{freqvar} = 1.668 \cdot 10^3$$

$$\text{freqstdev} := \sqrt{\text{freqvar}} \qquad\qquad \text{freqstdev} = 40.835$$

Figure 12

The majority of the data then falls within the range (49.6 - 40.8) and (49.6 + 40.8).

Explorations

The Basics

1. The following 3 data sets all have the same arithmetic mean of 50. For which of the data sets does the mean offer an accurate representation of the distribution of the data. Be quantitative in your analysis rather than qualitative.
   a) 0, 100, 25, 75
   b) 43, 50, 52, 55
   c) 20, 40, 60, 80

2. Typographical errors have different effects on the mean and median. Explore the effect of input errors on the data set {2, 4, 6, 8, 10, 12, 14, 16}. Is one measure more resilient than the other?

3. Measurements of the outside diameter of a hollow metallic cylinder were taken for a sample of 10 units from a production of 1000. The results are (in cm):
   10.25, 10.29, 10.22, 10.23, 10.23, 10.24, 10.20, 10.15, 10.32, 10.22
   a) Determine the minimum value, maximum value, median and mean of the sample.
   b) Determine the standard deviation for the data set.
   c) Which data points are within one standard deviation of the mean.
   d) Which data points are outside one standard deviation.
   e) Based on the results of parts c) and d), should the industrial process involved in manufacturing this part be reviewed?

4. An analysis of a neighborhood reveals the following age data:

age range	0 - 19	20 - 39	40 - 59	60 - 79	80+
number	125	335	407	253	35

   a) Plot a frequency polygon and a histogram for this data.
   b) Determine the mean of the distribution.
   c) Determine the standard deviation of the distribution.

Beyond the Basics

1. Analyze the data set created by the vector definition $data_k = rnd(2 \cdot k)$ for $k = 0$ to 400. Include the mean, median, frequency polygon and histogram plots, and a measure of the variance and standard deviation for the frequency distribution.

2. If a large number of measurements are taken on a system, the distribution of the measurement approaches a normal distribution having a theoretical form:

$$f = \frac{1}{\sqrt{2\pi}} e^{-x^2/2}$$ where f is the percentage frequency of the variable x. The factor of

$\dfrac{1}{\sqrt{2\pi}}$ ensures the area under the curve is 1.00 or that the curve is normalized.

   a) Create a normal distribution with the number of points, $N = 400$ and x defined over a range of [-4, +4].
   b) Plot a frequency polygon for the distribution.
   c) Determine the mean, variance and standard deviation of the normal distribution.
   d) Indicate, along the x-axis, points within the distribution that are 1, 2 and 3 standard deviations away from the mean.
   e) Double the range in x by doubling N. Does the overall shape of the curve change from the display in part b)? As far as you can tell from the plots, does the total area under the curve change?

5.2 Linear Regression and Interpolation

The exercises within the previous section covered the basics of descriptive statistics. The overall process involves the steps of data collection, analysis and interpretation. Obviously if you have a preconceived notion of the results so strong as to affect your analysis and interpretation of the data, the data was collected for the sole purpose of realizing your personal vision. If you assume the form of your data you may be imposing a bias, making the data fit your preferred view of the system's experimental behavior.

If, however, you approach your measurements in an objective manner then the numbers should reveal the system behavior. The analysis performed on the data should uncover an underlying pattern if one exists. Based on the analysis and its results, you can then interpret the relation and use it (perhaps) to predict future occurrences and behavior.

How do you gain knowledge of data between our measured points? How do we extrapolate to a level of input beyond our experimental range? We could make intelligent guesses or we could develop analytic routines which would use what we do know to estimate what we don't know. We fill in the gaps of our incomplete knowledge. Or, we extend our knowledge beyond the range of experimental inputs.

In this set of explorations, we will analyze a set of data and estimate points between the measured points. And, by examining the degree to which the data is or is not linear we can develop a routine to fit the best straight line to the data, the line with the least amount of deviance from the data points.

The algebraic details of this method of least squares are covered in more detail in Appendix 2, Linear Regression Theory.

Warmup

We will use some of the tools developed within the previous section. In a perfect experimental setup, we could collect the data and store it in a database for further analysis. The data would represent a series of like measurements on the system over a suitable range. The number of samples would be necessarily finite (somewhere beyond the range of 2). However, in this Warmup we will break the data collection into a separate file and fill it with pseudo-experimental data. The 'collected' data will represent a perfectly linear set of data with superimposed random error or noise. The data is then read into a series of unstructured text files for later retrieval and analysis.

Then the data will be read into the analysis file where it will be plotted and various statistical tools will be applied to it so as to determine its functional relationship.

Figure 1 shows the data creation process for the relation $y(x) = 5x + 10$. Noise is added to the data using the Mathcad random number operator, $rnd(x)$. This operator returns a random number from 0 to x. The offset ensures that the errors are applied both above and below the perfectly linear data.

$$N := 20 \qquad \text{number of samples}$$
$$k := 0 .. N \qquad \text{vector index}$$
$$X_k := k \qquad Y_k := 5.00 X_k + 10 \qquad \text{linear relation as vectors}$$

$$YRND_k := \left(Y_k\right) + rnd\left(4 \cdot X_k\right) - 0.5 \cdot rnd\left(4 \cdot X_k\right) \qquad \text{relation with added random error}$$

$$WRITE(\text{randsize}) := N \qquad \text{number of data elements to file}$$
$$WRITE(\text{randndex}) := X_k \qquad \text{x-axis data to file}$$
$$WRITE(\text{randdata}) := YRND_k \qquad \text{randomized y-axis data to file}$$

Figure 1

The relevant data has been written to a variety of *.dat files stored in Mathcad's default directory. You can convince yourself that the data has indeed been written to a file and has not ended up in some information black hole by opening any of the files in a text editor (e.g., Windows' Notepad, DOS's Edit).

Your data files now constitute a database, supposedly the result of repeated measurements on a system. These measurements may have been done by hand, the data noted in a laboratory book and later written (tediously) into a text file or, better yet, may have been input directly from the experimental apparatus without a middleman.

Now, we need to retrieve the information stored in our database. The information must be read and associated with a variable name. Figure 2 illustrates this process for the data generated in Figure 1 and includes a plot of the data. Note the scatter of the points about a seemingly linear relation.

$$DATASIZE := READ(\text{randsize}) \qquad \text{\# of samples read from file}$$
$$k := 0 .. DATASIZE \qquad DATASIZE = 20$$

$$XDATA_k := READ(\text{randndex}) \qquad \text{x-data read from file}$$
$$YDATA_k := READ(\text{randdata}) \qquad \text{y-data read from file}$$

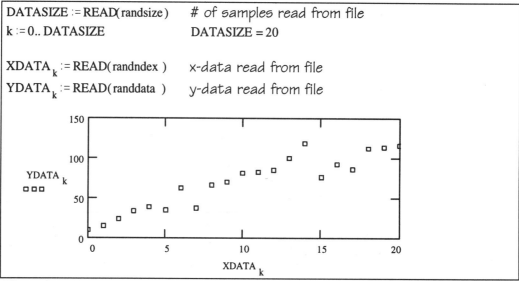

Figure 2

The data in the above plot represents our knowledge of the system, the extent of our measurements. These may, by themselves, be uncertain due to errors in measurements or the precision of the measuring instrument.

A simple approach to predicting the value of points between these data points involves connecting the known points with straight lines. This is a crude approach (connect-the-dots) but works sufficiently well if there are enough data points.

Mathcad offers the **linterp**(x-vector, y-vector, x-point) operator where x-vector and y-vector are the names of the respective *x* and *y* variables and x-point is the input value for which an output estimate will be generated. Figure 3 shows the application of the linterp operator to the previously generated data for values of *x* between the data input values.

$x := 0.5, 3.5 .. 18.5$	x	linterp (XDATA , YDATA , x)
	0.5	12.309
	3.5	36.383
	6.5	50.051
	9.5	75.568
	12.5	92.181
	15.5	84.077
	18.5	112.884

Figure 3

These points *between the points* are rough estimates only. The data points may indeed be linearly related but do not fall on an overall linear relation. And, the process does not take into account the possible non-linearity of the data.

For points beyond the range of the data, the **linterp** function is increasingly inaccurate. For a point above the range, the last two data points are used and a straight line is extrapolated. If the input point is not too far from the data, then a rough estimate can be determined. However, as the distance increases, so does the lack of reliability of the result.

Other interpolation tools do exist within Mathcad. However, they are beyond the range of this book.

Now that we have written and read and plotted the data, have found points in between the measured points, we can apply the linear regression tools. These will provide estimates of the best straight line through the scatter of data points. A best-slope and best-intercept will be generated and a measure given of the linearity of the data.

The routines automatically calculate the slope and intercept of the straight line which minimizes the overall squared vertical distance of the *y* data value from the best line. The square of the distance is used so that absolute distance rather than displacement is examined.

The form of the linear regression tools is **Operator**(x-data, y-data) where **Operator** is any of **slope** (for the best-fit slope), **intercept** (for the corresponding y-intercept) and **corr** (for the linear correlation coefficient, a measure between 0 and 1 or 0 and ÷1 of the linearity of the data). The application of these operators and the resulting straight line are shown in Figure 4. Note that the superimposed straight line passes through an average of the scatter of points. The correlation coefficient indicates that the data has a high probability of being linear. A perfectly linear data set would score a 1.00 while a perfectly random set would score a 0.

In this figure, the operators were used to define the variables best_m and best_b which were then used to define the best_fit straight line. The input into the straight line function was the set of original data points along the *x*-axis.

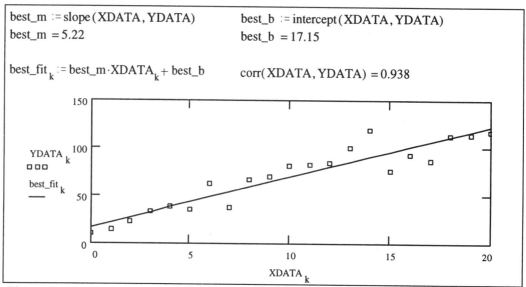

$best\_m := slope(XDATA, YDATA)$ $best\_b := intercept(XDATA, YDATA)$

$best\_m = 5.22$ $best\_b = 17.15$

$best\_fit_k := best\_m \cdot XDATA_k + best\_b$ $corr(XDATA, YDATA) = 0.938$

Figure 4

This new line can then be used to generate estimates of interpolated and extrapolated values with far greater confidence. Figure 5 illustrates the output of the best_fit line for input values outside the data set.

$j := 0..5$ for x-data, manually input as vector

$x\_out_j :=$ $best\_fit_j := best\_m \cdot x\_out_j + best\_b$ $best\_fit_j$

x_out_j		$best\_fit_j$
0.25		18.456
2.53		30.358
10.25		70.659
17.23		107.097
31.95		183.94
55.20		305.313

Figure 5

Explorations

The Basics

1. For the linear relation $y(x) = -4.0\,x + 10$, apply the processes of linear interpolation and linear regression. Rather than interpret the relation as a function, use the model of a vector relation developed within the Warmup section. What does the negative correlation coefficient signify?

2. Before Ohm's Law was a law, G.S. Ohm had to make measurements of voltage and current across a resistive element. The slope of the line generated from data of current (I) versus voltage (V), with voltage as the independent variable, provided an estimate of the reciprocal resistance (the conductance). Use the linear regression and interpolation tools to determine the resistance for the following I-V data set:

Voltage (V)	1.00	4.00	9.00	16.00	25.00
Current (A)	0.0044	0.0163	0.0321	0.0663	0.1102

3. Use your analysis in #2 to predict values of the current for voltages of
     a) 0.25 Volts
     b) 35.2 Volts
4. Apply the process of linear regression to the quadratic equation $y(x) = 3x^2 + 5$ for x on the domain [0,10]. You should, amazingly enough, produce a straight-line fit with a high degree of correlation. Alter the coefficient of x^2 for the same domain. As the coefficient varies, at what point does the linear plot not correspond to the quadratic data?
5. Repeat exercise #4 for a general cubic function $y(x) = ax^2 + bx + c$. Does the degree of correlation depend:
     a) on the number of vector points?
     b) on the values of a, b and c?
     c) on the range in the values of x?

Beyond the Basics

Linear regression can be applied to supposedly non-linear data. If the data fails to be linear then it may have a power function form $y(x) = ax^n$ or an exponential form $y(x) = a\,b^x$.

In the case of a power function $y = ax^n$ the application of the 'log' or 'ln' transforms the relation into $\log(y) = \log(a) + n\log(x)$. This relation has a linear form with the slope defined as n, the y-intercept as $\log(a)$. A linear regression can be applied to the data of $\log(x)$ and $\log(y)$, values of n and a can be determined from the best-fit slope and intercept respectively and the original function can be reconstructed.

For an exponential function, the same method applies. The application of the 'log' transforms the relation into $\log(y) = \log(a) + x\log(b)$. The regression process can then be applied to the data set of $\log(y)$ and x. The intercept is represented by $\log(a)$ while the slope is given by $\log(b)$.

1. Generate a data set for the general relation $y(x) = a\,x^n$ for values of a, n and the domain of your choice. Apply the linear regression routine to the raw data (x and y) and determine the best-fit line and the correlation coefficient. Then, re-express the power function in terms of its log (or ln) transformation and apply the regression routine to this linear expression. What value of the correlation coefficient do you expect from the transformed data? Do the values of a and n generated from the best-fit applied to the transformed function correspond to the original values you selected?
2. The following data set represents a power function with added error. Determine the function which best describes the data set's behavior.

x-data	0	1	2	3	4	5	6
y-data	0	5.4	83.9	400.0	1299.2	3025.0	6500.0

3. The following data set represents an exponential function with added variance. Determine the function which best describes the data set's behavior.

x-data	0	1	2	3	4	5	6
y-data	3.1	8.8	21.2	64.3	160.5	445.2	1200.0

Chapter 6: Trigonometric Functions

Trigonometry, the study of the angles and sides of the triangle, is one of the great hidden tools within our technologies. It was the premier tool of ancient calendar makers and priests and of the architect-builders. The old practitioners have been dust for centuries but their creations live on.

Trigonometry allowed the first estimates of the distances from the earth to the moon and from the earth to the sun. It was used to approximate the circumference of the earth based on the noontime shadow of a vertical marker centuries before European culture (re-)discovered the Earth as a sphere.

We may think the value of trigonometry has become obscured with time. After all, it is no longer necessary to determine the height of the opponent's cannon above the valley floor as we prepare for an attack on the battlement. Our calendars are well fixed and can be printed years ahead of time (some with pictures of cute animals, no less). The phases of the moon and the position of the sun in the heavens can be accurately predicted. The dates of religious holidays and solar eclipses are known well in advance. And most of our buildings are plumb and level.

As with many other areas of mathematics, the knowledge has become implicit in our culture, hidden under layers and layers of technology. If you dig into the areas of telecommunications, linear circuit and signal analysis, mechanical design and stress analysis, electrical and civil engineering, terrestrial and celestial navigation, and countless other technological fields, you will find trigonometry.

The tool has indeed survived.

6.1 Trigonometric Functions, Their Reciprocals and Inverses

Definitions of the trigonometric functions are just that: definitions. For a constant angle, the ratios of the sides of the right triangle do not vary as the sides are scaled up and down. A triangle with sides in the ratio 3:4:5 scaled to 6:8:10 exhibits exactly the same trigonometric relations.

The three basic trigonometric functions of sine, cosine and tangent are essentially interconnected. The cosine of an angle is the sine of its complement. As a function, the cosine is simply a sine curve shifted along the horizontal axis. The tangent can be defined as the ratio of the sine to the cosine.

We will examine the units of angular measurement, the numerical and symbolic evaluation of the basic trigonometric functions, the evaluation of the reciprocal functions and the inverse functions.

Warmup

Trigonometric Ratios and Their Measure

The angle between two intersecting lines can be viewed as the degree of rotation necessary at the intersection point that would rotate the one line onto the other.

In the same way that other physical units are based on arbitrary measurements, the ancient Babylonian system of 60s has survived as a system of angular measurement. A full rotation of a line about the origin covers 6 of these groups of 60. We have come to associate the circle with 360° rotation. As a line segment rotates about the origin, it sweeps over 360 degrees of arc. However, there is nothing sacred (at least nowadays) with the measure. We could as easily have defined the circle as having 400 units (as in the case of the engineering unit of gradient) or 10 units (a decimal circle).

However, one system exists which is independent of the observer's whim. The radian, defined as the angle subtended by an arc equal to the circle's radius, is a unit dependent only on the definition of the circular circumference as $C = 2\pi r$. The number of radians in one revolution about the origin is then the ratio of the circumference to the arc length or 2π.

Mathcad's default angular measure is the radian. All outputs of the operation of sine, cosine, or tangent on an argument are assumed to have operated on a radian measure of angle. However, many people prefer their trigonometric input in degrees. Within the worksheet, you can define custom trigonometric functions which accept degree measure or, in the case of the output of the inverse functions, multiply the default output of radians by the unit 'deg'.

Figure 1 shows tables of values for a range of angles we wish to have interpreted as being in degrees. The conversion factor R allows one (not very elegant) way of translating the angle in degrees to a radian input. A less intrusive method simply attaches the unit 'deg' to the placeholder attached to *x*. The expression 'x·deg' generates a table of angle values in radians.

☑ This may provide a point of confusion as most units simply define the scalar they are attached to rather than convert it to another system of measurement.

$$x := 0, 10 .. 50 \qquad R := \frac{\pi}{180}$$

x	R·x	sin(R·x)	cos(R·x)	tan(R·x)	x·deg	sin(x·deg)
0	0	0	1	0	0	0
10	0.175	0.174	0.985	0.176	0.175	0.174
20	0.349	0.342	0.94	0.364	0.349	0.342
30	0.524	0.5	0.866	0.577	0.524	0.5
40	0.698	0.643	0.766	0.839	0.698	0.643
50	0.873	0.766	0.643	1.192	0.873	0.766

Figure 1

Note that the columns for sin (R · x) and sin (x · deg) output exactly the same values for the range of angles from 0° to 50°. We can as easily define degree-measure trigonometric functions which by definition accept degree input. Figure 2 illustrates the definitions and the corresponding tables of values.

$$sind(x) := sin(x \cdot deg) \qquad cosd(x) := cos(x \cdot deg) \qquad tand(x) := tan(x \cdot deg)$$

$$x := 0, 10 .. 50$$

x	sind (x)	cosd (x)	tand (x)
0	0	1.00000	0
10	0.17365	0.98481	0.17633
20	0.34202	0.93969	0.36397
30	0.50000	0.86603	0.57735
40	0.64279	0.76604	0.83910
50	0.76604	0.64279	1.19175

Figure 2

Plots of the Trigonometric Functions

As the hypotenuse of the right triangle rotates about the origin in the counter-clockwise (ccw) direction, it covers a range of angles measured relative to the positive *x*-axis. If the hypotenuse is of unit length, the sine function maps out the length of the side opposite the angle, the cosine function maps out the length of the adjacent side and the tangent keeps tracks of the ratio of (opposite/adjacent).

Figure 3 shows the functions sin(*x*) and cos(*x*) over the domain [0, 3π]. As the hypotenuse rotates through the first quadrant and the angle increases from 0 to π/2 radians, the length of the opposite side (defined by the sine function) grows while that of the adjacent (defined by the cosine) decreases.

Note that the functions of sine and cosine resemble one another except for a shift of π/2 along the *x*-axis. This relation is expected as they both track the movement of the sides of the right triangle as the hypotenuse rotates. And the side adjacent to an angle θ is the side opposite its complementary angle (π/2 - θ) for a right triangle. This last property is the origin of the co-function relation between the sine and its co-sine.

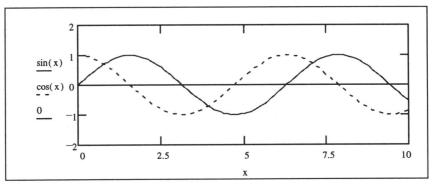

Figure 3

Figure 4 shows the sine and cosine functions with angles of π/4 and 5π/4 marked and with the horizontal scale in units of π/2 radians. The tangent function, defined as the ratio of the sine to the cosine, has a value of 1 at $x = π/4$. At multiples of π, where the sine is zero and the cosine 1 or -1, the ratio is zero.

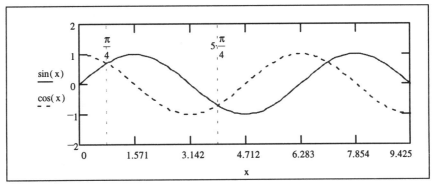

Figure 4

Thus from 0 to π/4 the tangent function grows from 0 to 1. If you picture the marker as a moving vertical line, then as the marker moves right beyond π/4, the sine function grows to 1 while the cosine decreases to zero. The tangent's behavior must then explode to infinity and be undefined at the angle π/2. From the definition of the tangent as a ratio of sides, this behavior makes sense as the adjacent side collapses to a length of zero as the angle approaches π/2.

As the marker passes through π/2, the value of the cosine function moves to small negative values while the sine function is still in its maximum region. The tangent must then skip to high negative values over this transition. The result is a discontinuous function at odd multiples of π/2.

Tables of values of sine, cosine and tangent in the region of 90° (π/2 radians) are provided in Figure 5. The figure shows that the definition of the tangent in terms of sine and cosine is consistent.

x	sind (x)	cosd (x)	$\dfrac{sind\,(x)}{cosd\,(x)}$	tand (x)
89.15	0.9998900	0.0148348	67.402	67.402
89.30	0.9999254	0.0122170	81.847	81.847
89.45	0.9999539	0.0095992	104.171	104.171
89.60	0.9999756	0.0069813	143.237	143.237
89.75	0.9999905	0.0043633	229.182	229.182
89.90	0.9999985	0.0017453	572.957	572.957
90.05	0.9999996	- 0.0008727	- 1145.915	- 1145.915
90.20	0.9999939	- 0.0034907	- 286.478	- 286.478
90.35	0.9999813	- 0.0061086	- 163.700	- 163.700
90.50	0.9999619	- 0.0087265	- 114.589	- 114.589
90.65	0.9999357	- 0.0113444	- 88.144	- 88.144
90.80	0.9999025	- 0.0139622	- 71.615	- 71.615

Figure 5

Figure 6 shows a plot of the tangent function, tan(*x*). Although the discontinuities are evident in Figure 6(a), the pattern Mathcad presents does not appear to be too helpful. In fact, the behavior of the function is being obscured by the very tool which should help to illuminate it. A change of vertical scale, as presented in Figure 6(b), helps clear up some of the graphing problem. However, since the plotting is accomplished point-to-point, there is a tendency to include a line connecting the infinities. This line is an artifact, a remnant left over from the plotting process, and not an indication of the true behavior of the function.

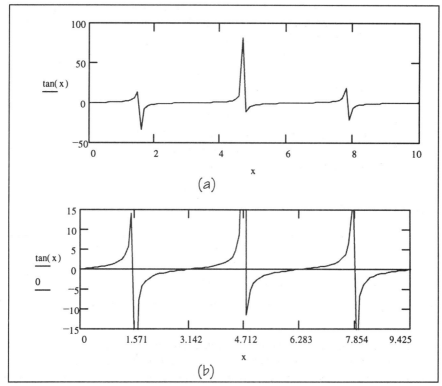

Figure 6

Symbolic Calculations

Mathcad's symbolic process allows the exact evaluation of trigonometric ratios if such closed forms exist. Figure 7 illustrates the symbolic engine applied to the regular trigonometric functions. From within the menu, Symbolic/Evaluate/Evaluate Symbolically was selected. The output expressions were then evaluated numerically by selecting Symbolic/Evaluate/Floating Point Evaluation with a precision of 5.

☑ In Version 5.0, load the symbolic engine (select Symbolic/Load Symbolic Processor) before proceeding with the calculations in Figure 7.

(a)	(b)	(c)	(d)	(e)	(f)
$\sin\left(\dfrac{\pi}{6}\right)$	$\cos\left(\dfrac{\pi}{3}\right)$	$\sin\left(\dfrac{\pi}{3}\right)$	$\sin\left(\dfrac{\pi}{12}\right)$	$\sin\left(3\cdot\dfrac{\pi}{8}\right)$	$\tan\left(\dfrac{\pi}{6}\right)$
exact symbolic value					
$\dfrac{1}{2}$	$\dfrac{1}{2}$	$\dfrac{1}{2}\cdot\sqrt{3}$	$\dfrac{1}{2}\cdot\sqrt{2-\sqrt{3}}$	$\dfrac{1}{2}\cdot\sqrt{2+\sqrt{2}}$	$\dfrac{1}{3}\cdot\sqrt{3}$
floating point evaluation - precision of 5					
.5	.5	.86605	.2588	.9239	.57736

Figure 7

Reciprocal Functions

The three basic trigonometric functions of sine, cosine and tangent have their reciprocals defined as separate functions. These reciprocal functions should not to be confused with the inverse trigonometric functions. The reciprocal of a function $f(x)$ is defined as $1/f(x)$ while the inverse of a function $f^{-1}(x)$ is the operation that undoes the original action of the function. As an example, while the reciprocal of the sine function is the cosecant function defined as $1/\sin(x)$, the inverse sine function takes the ratio initially output by the sine function and returns the angle (or its equivalent).

These reciprocal functions are built-in to Mathcad as shown in Figure 8. The use of these functions is preferred rather than risk the confusion associated with assigning the power of -1 to the trigonometric function.

$\cos(3.5) = {}^-0.936$	$\sec(3.5) = {}^-1.068$	$\dfrac{1}{\cos(3.5)} = {}^-1.068$
$\sin(4.0) = {}^-0.757$	$\csc(4.0) = {}^-1.321$	$\dfrac{1}{\sin(4.0)} = {}^-1.321$
$\tan(12) = {}^-0.636$	$\cot(12) = {}^-1.573$	$\dfrac{1}{\tan(12)} = {}^-1.573$

Figure 8

Plots of the reciprocal functions are given in Figure 9. Discontinuities exist for the reciprocals of the sine and cosine functions where these have values of zero. In the case of the tangent function, its reciprocal moves to zero as its own value moves toward the infinities and vice versa.

Note the co-relation between the following pairs of functions: sine-cosine, secant-cosecant, tangent-cotangent. The output of the one function within the pair is equal to the output of the co-function applied to the angle's complement. The relation is function (x) = co-function $(\pi/2 - x)$.

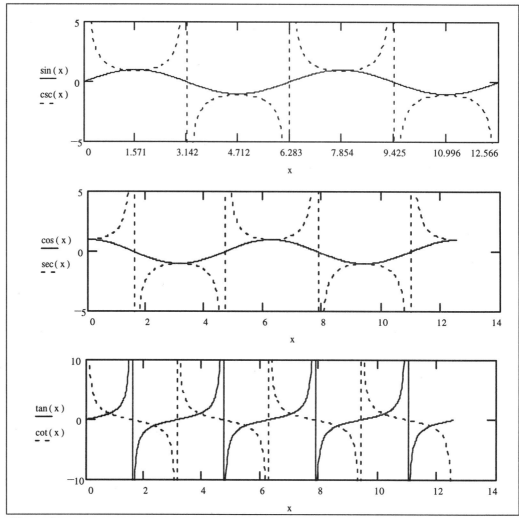

Figure 9

Inverse Functions

The inverse functions return an angle for the input of a particular ratio. Although the trigonometric functions and their reciprocals are single-valued (one output for each input), the inverse of a trigonometric ratio is multi-valued. An infinite number of possibilities exists for each particular ratio. This is the effect of the cyclical behavior of the sine, cosine and tangent and is illustrated in Figure 9 for the equation $\sin(x) = 0.707$.

Figure 9

As a result of this bounty of possibilities, the built-in functions of asin(x), acos(x) and atan(x), named for the arcsine, arccosine and arctangent functions respectively, are defined only over principal values of the output angle. Figure 10 shows the inverse functions for the sine and cosine while Figure 11 shows the arctangent function.

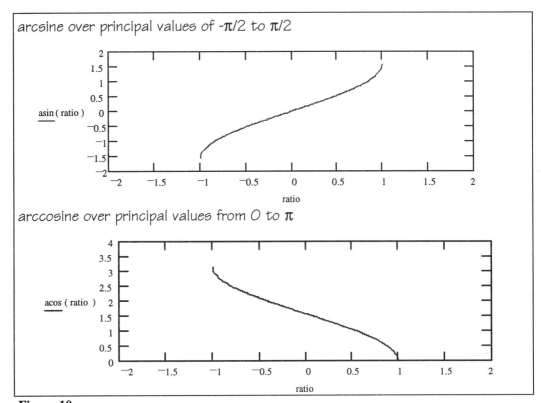

Figure 10

☑ As can be seen from the above plots, the returned angle exists in only half of the plane. If, instead, the equivalent angle in the other half of the plane had to be returned, its value can be determined using the CAST rule.

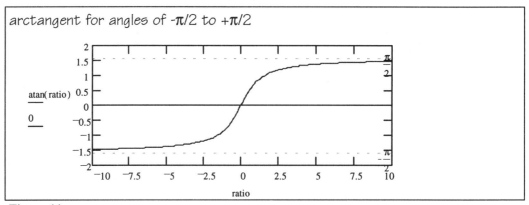

Figure 11

Generalized Trigonometric Functions

So far the form of the functions we have been examining has been simple. The input angle has been fed directly into the trigonometric operator and an output has been produced.

The most general form of the sine function is $y(x) = A \sin (B x + C)$. This same generalized form can be applied to the five other trigonometric functions. The meaning of the constants and coefficient will be examined within the Explorations.

Explorations

The Basics

1. By defining $y(x) = A \sin(x)$, examine the effects of changes in the constant A for the generalized form of the sine function. Are there any restrictions to the types of numbers A can be?
2. For the function $y(x) = \sin(Bx)$, what is the effect of changes in the coefficient B? What does a negative value of B do to the sinusoidal pattern? How would you define the overall effect of this coefficient?
3. Apply the constant C to the function $y(x) = \sin(x)$ so that the new function is defined as $y(x) = \sin(x + C)$. What is the effect of this constant on the pattern of the sine wave? At which values of C does the sine function exactly copy the cosine function?
4. From the results of #3, you should expect a well defined shift in the pattern of the basic sine function for the addition of the constant C while the frequency is set to 1. Now set $B = 2$ with $C = \pi/4$. Does the shift in the sine pattern differ from the one you expected?
5. Generate any sinusoidal function by defining A, B and C. Predict the visual pattern (amplitude, frequency, phase angle, phase shift) before plotting the function.
6. Repeat *The Basics* #1 to # 5 for the cosine or tangent functions.
7. What is the effect of adding a constant to the trigonometric function itself so that you generate $y(x) = \text{sinusoid}(x) + \text{constant}$?

Beyond the Basics

1. Show, using tables and graphs, that the relation, function (x) = co-function $(\pi/2 - x)$, holds for each of the sine-cosine, secant-cosecant and tangent-cotangent pairs.

2. Trigonometric functions can be superimposed to yield unexpected patterns. Examine the following functions:

   a) $y(x) = 3\sin(3x) + 2x$ 　　　　 b) $y(x) = \sin(x) + \cos(2x)$

   c) $y(x) = \sin(2.5x) + \sin(3x)$ 　　 d) $y(x) = \sin(x) + \dfrac{1}{3}\sin(3x) + \dfrac{1}{5}\sin(5x)$

3. Develop a one page Mathcad file which will be used as a sine function tutor by another. Define a general function and invite the viewer to edit your file so that the effects of the various parts of the generalized sine function can be examined.

4. The sine function has so far accepted a value of the angle and output an amplitude. The most common domain of the sinusoidal functions is the time domain. The generalized expression for the sine wave as a function of time is $y(t) = A\sin(\omega t + \phi)$ where $\omega = 2\pi f$. The frequency f is defined as the number of cycles per second and is measured in Hertz (Hz). You can visualize a hypotenuse of length A rotating about the origin at a rate of f cycles/second. This hypotenuse sweeps out $2\pi f$ radians per second of angle. The phase angle ϕ represents the direction of the rotating hypotenuse at the time $t = 0.0$ s. The phase shift, given by $-(\phi/\omega)$, represents the time by which the hypotenuse leads or lags the reference wave of the same frequency but with zero phase angle. Figure 12 illustrates the time varying sine wave.

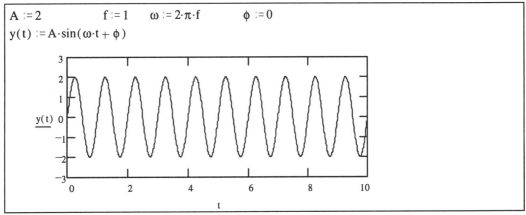

Figure 12

Plot the function $y(t) = A\sin(\omega t + \phi)$ for various selections of A, ω, and ϕ. Try to keep the other variables constant so that you can determine the effect of each of the variations.

5. The waves (in angle or in time) have so far been static. In the previous exercise, the plot shows the amplitude of the function as it passes through a particular point in space. Construct a wave which travels in time (presents shifted waves for the plot of y versus x or y versus t) by using the definition $y(t, x) = A\sin(kx - \omega t)$ and varying both x and t at the same time over suitable ranges. The constant k represents $2\pi/\lambda$ where λ is the wavelength of the traveling wave.

6.2 Trigonometric Equations

The previous set of explorations covered the definitions of the basic trigonometric functions and emphasized the visualization of the parts of the generalized functions. Part of the problem in the inverse functions is the infinite number of possibilities they allow due to the cyclical nature of the trigonometric functions.

For a trigonometric equation, this infinity creates a problem. As a simple example, the equation $\sin(x) = 0.5$ would have solutions for x at $\pi/6 + 2\pi k$ and $5\pi/6 + 2\pi k$ where k is an integer. Clearly, from a practical point of view, only the solutions for $x = [0, 2\pi)$ are necessary.

In this set of exercises we will examine analytic, graphical and numerical methods in the solution of trigonometric equations.

Warmup

For the solution of a simple sinusoidal equation, we will use tools we are already familiar with. Plots give an overall view of the function. The symbolic processor can be used to output symbolic or analytic solutions while we have examined a numeric process, the Solve Block that will allow us to examine solutions not amenable to symbolic interpretation.

For the equation $\sin(x) = 0.5$, a plot of the two functions $y(x) = \sin(x)$ and $y(x) = 0.5$ in Figure 1 shows 2 intersection points over the domain $[0, 2\pi)$, 2 points at which the equation is satisfied. The intersection pattern is then repeated over $[2\pi, 4\pi)$.

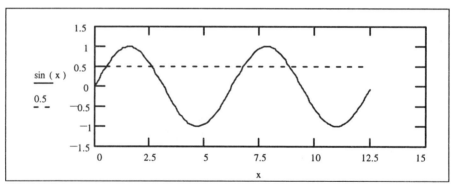

Figure 1

The Symbolic Processor can be applied to the equation by selecting the variable x and then using Symbolic/Solve for Variable. Figure 2 shows the output of the application of the Symbolic Processor. Note that counter to the multiplicity of available answers, only one answer is returned.

$\sin(x)=0.5$	has a solution for x of	.523598775598298873
compared to asin(0.5)	$asin(0.5) = 0.5235987756$	

Figure 2

We can however use a numeric solution process with our plot as a guide. From Chapter 3, Section 2, the Solve Block routine is used to solve equations numerically based on a seed point. By changing this point of origin, the routine allows multiple solution points. We first examine the plots generated from treating each side of the equation as a separate function of x from which an approximate point can be defined. As Figure 3 shows, this becomes the starting point for the numerical process and is used in defining the separate solutions.

a := 0.5	b := 2.0
Given	Given
sin(a)≈0.5	sin(b)≈0.5
Solutiona := Find(a)	Solutionb := Find(b)
Solutiona = 0.524	Solutionb = 2.618

Figure 3

Any other solution to the equation can be examined by choosing an appropriate approximation point.

Explorations

The Basics

1. Solve the equation $2\sin(3x) - 2 = -x$ using symbolic and numerical techniques. The plot of the separate functions $y(x) = 2\sin(3x) - 2$ and $y(x) = x$ is given in Figure 4.

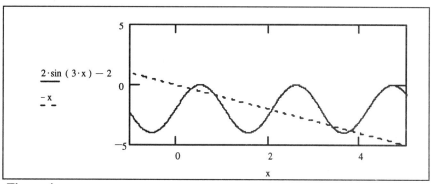

Figure 4

2. The equation in #1 can be rewritten as $2\sin(3x) - 2 + x = 0$. Although the plot may appear different than that given in Figure 4, does this reorganization of the original equation alter the solution(s) in any way?
3. The equation $\cos(x) = \sin(3x)$ has six distinct solutions over [-3, +3]. The behavior of these intersection points is then repeated for values of x beyond this domain. Determine these six values using symbolic processing, graphical estimates and the Solve Block routine.
4. Solve the equation $\sin^2(2x) = 0.5$ over $[0, 2\pi]$.

Beyond the Basics

1. Show that the equation $2\cos(x) - 1 = 2$ has no real solution(s). However, the symbolic solver will return an answer. What is the meaning of this complex solution?
2. The equation $6\cos^2(x) = 5\cos(x) + 4$ can be solved as a quadratic equation of the form $ay^2 + by + c = 0$ with $y = \cos(x)$. Find all solutions of the angle x between 0 and 2π.

Chapter 7: Complex Numbers

The introduction of complex numbers is often viewed with some dismay. The names associated with the parts of a complex number, real and imaginary, seem to suggest the dominance of the first part over the second. After all, a physical measurement is a real number. An imaginary number must be somehow of less value by virtue of its un-real-ness.

Defined as $\sqrt{-1}$, the imaginary number i or j is seen as a useful construction, an answer to the problem of defining solutions to equations such as $x^2 = -1$ or to quadratic equations with discriminants less than zero. The pure study of complex numbers does not require any further definition or explanation. A consistent and fascinating area of mathematics emerges if one merely accepts the definition and then proceeds to examine the properties of such numbers.

In an applied sense, however, an interpretation of the imaginary number which is more useful is that of being one part of a dimensional pair, the Imaginary axis rotated 90° with respect to the Real. The symbol j then takes on the more physically real identity of a 90° rotation operator. As j operates on a complex number, it effects change. It rotates the complex number (represented as a vector) 90° about the origin. Repeated applications of the j operator can then account for continued cyclical rotation clockwise or counter clockwise about the origin.

In this set of explorations, we will examine the nature of the imaginary operator, the arithmetic of complex numbers and the application to time varying systems.

7.1 The Use of Complex Numbers

The power of the complex notation as applied to technology lies in its algebra. In time-varying systems, the algebra of complex numbers is the same as the algebra of the time-varying components (e.g., voltages and currents in AC circuit theory). The representation, using complex numbers, of sinusoidal fields governing the propagation of the electric and magnetic fields allows an especially elegant and compact form of the equations and their manipulations.

The multiplication and division of these quantities can be accomplished using traditional trigonometric identities. However, complex numbers allow a clearer description of their properties. This last statement is not a paradox but an unfortunate and enduring use of the word 'complex'.

Warmup

Mathcad allows the representation of the imaginary component using either the *i* or *j* notation. While engineers prefer the *j* notation (possibly to lessen its confusion with the symbol *i* for electric current), rigorous mathematicians prefer the *i* notation. Both factions can agree on its usefulness, whatever the symbol.

To alter the default symbol for the imaginary operator, choose the preferred option from the Math/Numerical Format dialog box. Make sure that the format you choose is recognized as a Global formatting option.

Although you have now chosen a preference, you must also define the symbol in any documents using complex numbers. Figure 1 shows the definition of the symbol *j* (with the squareroot sign defined by '\') and its use within a complex number, *z*.

$$
\begin{array}{ll}
j := \sqrt{-1} & \text{definition of } j \text{ operator} \\
x := 3 \quad y := -2 & \text{real and imaginary parts} \\
z := x + y \cdot j & \text{definition of complex number} \\
z = 3 - 2j & z\text{'s value in complex plane} \\
|z| = 3.606 & \text{amplitude of } z \\
\arg(z) = -33.69 \cdot \deg & \text{argument of } z \text{ (degrees)} \\
\arg(z) = -0.588 & \text{argument of } z \text{ (radians)}
\end{array}
$$

Figure 1

☑ Care must be taken with the use of the *j* symbol. Mathcad would try to recognize an isolated expression in *j* (e.g. *jx*, *xj*) as a variable name. And, while Mathcad recognizes the expressions *j*, 1*j*, and 2*j* as valid, it will return an error message for the expressions *j*1 or *j*2. It is probably better to include the multiplication symbol within all operations involving *j* and make its use explicit.

The argument of the complex number *z* is the angle its vector representation makes with the positive real axis and is invoked by typing '*arg(z) =*' . An angle in radians is returned from $+\pi$ to $-\pi$. The magnitude of *z* (or length of the hypotenuse of the triangle) is determined by selecting |x| from the vertical palette (V5.0) or from the Arithmetic Palette (V6.0) or by typing '| *z* ='.

The default angular measure is radians but the use of degrees may be more appropriate as the degree measure is sometimes easier to visualize. Figure 1 includes the same argument in both unit systems. To change the output radian measure to degrees, select the unit place holder to the right of the angle output and type 'deg'. Upon recalculation, Mathcad will update the angular measure to the equivalent number of degrees.

Now that the basic Mathcad functions of magnitude and argument have been applied to the complex number, we may examine the way in which complex arithmetic is handled.

Addition and subtraction occur in a real to real and imaginary to imaginary fashion following the arithmetic of vectors. In a 2-dimensional plane, the real and imaginary components (x and y respectively) form the orthogonal (90°) components of the 2-dimensional vector z. Figure 2 shows the definition of two complex numbers and samples of their addition, subtraction and scalar multiplication. The results are just those you would expect from vector analysis.

Note that the argument of z_1 is unchanged upon scalar multiplication ($3 \cdot z_1$). The length of the vector has been increased by a factor of 3 but its direction remains unchanged.

$$z_1 := 4 + 5 \cdot j \qquad z_2 := 3 - j \cdot 7 \qquad \text{definition of two complex numbers}$$

$$|z_1| = 6.403 \qquad\qquad |z_2| = 7.616$$
$$\arg(z_1) = 51.34 \cdot \deg \qquad \arg(z_2) = -66.801 \cdot \deg$$

$$z_1 + z_2 = 7 - 2j \qquad z_1 - z_2 = 1 + 12j \qquad \text{addition/subtraction}$$

$$3 \cdot z_1 = 12 + 15j \qquad \arg(3 \cdot z_1) = 51.34 \cdot \deg \qquad \text{scalar multiplication}$$
$$4 \cdot z_1 - 2 \cdot z_2 = 10 + 34j$$

Figure 2

If physical quantities such as voltage and current can be represented by these complex vectors, how is the multiplication and division of these quantities to be interpreted (and determined)?

Addition and subtraction seem to be reasonable and straightforward. With multiplication and division, the true (to engineers) nature of the j symbol appears.

If we return to our original vector $z = 3 - 2j$, we find its magnitude to be 3.606 and its argument (or angle) to be -33.69°. Figure 3 shows the result of repeated application of the j operator to the vector z as jz, $j^2 z$, $j^3 z$ and $j^4 z$.

$$
\begin{array}{lll}
z = 3 - 2j & |z| = 3.606 & \arg(z) = -33.69 \cdot \deg \\
j \cdot z = 2 + 3j & |j \cdot z| = 3.606 & \arg(j \cdot z) = 56.31 \cdot \deg \\
j^2 \cdot z = -3 + 2j & |j^2 \cdot z| = 3.606 & \arg(j^2 \cdot z) = 146.31 \cdot \deg \\
j^3 \cdot z = -2 - 3j & |j^3 \cdot z| = 3.606 & \arg(j^3 \cdot z) = -123.69 \cdot \deg \\
j^4 \cdot z = 3 - 2j & |j^4 \cdot z| = 3.606 & \arg(j^4 \cdot z) = -33.69 \cdot \deg
\end{array}
$$

Figure 3

In the first case, the j operator has maintained the magnitude of z but has rotated the vector +90° about the origin. The application of j to this new vector, $j\,(jz)$ or $j^2 z$, returns a vector +180° away from the original.

Herein lies the operator definition of j. Since the repeated application returns the negative of the original vector, then $j^2 z = -z$ or $j^2 = -1$. Likewise, $j^4 z$ is equivalent to four successive $+90°$ rotations and should bring the vector back to its original position. And it does.

How would $(j^{15} z)$ be interpreted? Or $[(-j)^3 z]$? Check that the results are consistent with the definition of a $+90°$ rotation operator.

The imaginary number $1j$ or $j1$ can be considered to be the number 1 along the positive real axis rotated by $+90°$ onto the positive imaginary axis. If the magnitude of the j rotator is increased to 3 then the total effect is to scale the vector by a factor of 3 and rotate it $+90°$. Multiplication of two complex numbers then combines scaling with multiplication. Figure 4 shows the multiplication of z_1 and z_2 using the numerical and symbolic tools in Mathcad.

$$z_1 = 4 + 5j \qquad |z_1| = 6.403 \qquad \arg(z_1) = 51.34 \cdot \deg$$
$$z_2 = 3 - 7j \qquad |z_2| = 7.616 \qquad \arg(z_2) = -66.801 \cdot \deg$$

using Symbolic Processor...

$$\left(x_1 + j \cdot y_1\right) \cdot \left(x_2 + j \cdot y_2\right) \quad \text{expands to} \quad x_1 \cdot x_2 + x_1 \cdot j \cdot y_2 + j \cdot y_1 \cdot x_2 + j^2 \cdot y_1 \cdot y_2$$
$$(4 + 5j) \cdot (3 - 7j) \qquad \text{expands to} \qquad 47 - 13j$$

using numerics

$$z_1 \cdot z_2 = 47 - 13j$$
$$|z_1 \cdot z_2| = 48.765 \qquad \arg(z_1 \cdot z_2) = -15.461 \cdot \deg$$
$$|z_1| \cdot |z_2| = 48.765 \qquad \arg(z_1) + \arg(z_2) = -15.461 \cdot \deg$$

Figure 4

In rectangular coordinates, the multiplication of two complex numbers follows the usual expansion rules as seen in Figure 4's use of the Symbolic Processor to expand the generalized forms of z_1 and z_2. However, the polar form (magnitude + argument) allows a more elegant method of determining the result and shows that the operation of multiplication includes a product of the two amplitudes and a sum of the angles. The process of multiplication creates vector rotation. Division must then undo the process of multiplication. The amplitudes are divided and the angles subtracted as demonstrated in Figure 5. The complex conjugate of a complex number is applied by selecting the expression then pressing the double-quote symbol.

Figure 5 also includes a description of complex division in the rectangular form. It involves multiplication of the numerator and denominator by the denominator's complex conjugate and should not be attempted by the faint of heart or the impatient.

$$z_1 = 4 + 5j \qquad z_2 = 3 - 7j \qquad \overline{z_2} = 3 + 7j \qquad z_1, z_2 \text{ and conjugate of } z_2$$

$$\frac{z_1}{z_2} = -0.397 + 0.741j \qquad \left|\frac{z_1}{z_2}\right| = 0.841 \qquad \arg\left(\frac{z_1}{z_2}\right) = 118.142\,\text{deg}$$

$$\frac{|z_1|}{|z_2|} = 0.841 \qquad \arg(z_1) - \arg(z_2) = 118.142\,\text{deg}$$

use of complex conjugate in division

$$z_1 \cdot \overline{z_2} = -23 + 43j \qquad \frac{z_1 \cdot \overline{z_2}}{z_2 \cdot \overline{z_2}} = -0.397 + 0.741j \qquad z_2 \cdot \overline{z_2} = 58$$

Figure 5

Explorations

The Basics

1. For the complex number, $z = -5 - j7$, determine the magnitude and argument. Are there other angles (arguments) which could be used to represent this complex vector?
2. For the complex number, $z = -6 - j9$, determine the rectangular and polar expressions for

       a) z^2 b) z^5 c) z^{12} d) $(-z)^3$

   How do the rectangular and polar representations of these results compare? Develop the general formula for a complex number raised to a power.
3. Perform the following complex number operations on the given vectors. Here, z^* represents the complex conjugate of z.

   $$z_1 = -3 + j2 \qquad z_2 = 6 + j3 \qquad z_3 = -1 - j4 \qquad z_4 = 5 - j3$$

     a) $\dfrac{z_2}{z_1}$ b) $\left(z_4\right)^8$

     c) $\left(z_1 \cdot z_2\right)^2$ d) $\left(z_2 \cdot z_4\right)^{1/2}$

     e) $3z_1 + 4z_3 - 2z_2 + z_4$ f) $\left(\dfrac{z_3}{z_4}\right)^3 \cdot z_2 + z_1$

     g) $j^3 \cdot \left(z_1 \cdot z_3\right)^2$ h) $\dfrac{z_1 + z_2}{\left(z_3 + z_4\right)^2}$

     i) $\left(\dfrac{z_3}{z_4^*}\right) \cdot z_1$ j) $\left(z_3 \cdot z_4\right)^*$

4. Show, for any complex number z, $\sqrt{z \cdot z^*} = |z|$.
5. Use the Symbolic Processor to solve the quadratic equation: $6x^2 + 2x + 9 = 0$. Show that no real roots exist using a plot and the Solve Block routine.

Beyond the Basics

1. Show, for all x, $e^{jx} = \cos(x) + j\sin(x)$. This relation, known as Euler's Formula, allows an extremely compact way of expressing and operating with complex numbers.
2. Use Euler's Formula to express the multiplication and division of two sinusoidal quantities $V_1 = 15.0\, e^{j\,10t}$ and $V_2 = 0.5\, e^{j\,6t}$. Compare the exponential result to the

result obtained from multiplying the equivalent expanded expressions in cosines and sines.

3. Determine the fifth root of the complex number $z = 7 - 2j$. Do other fifth roots exist? If so, how are they related to the principal root?

4. Solve the equation $z^4 = 1$ for all possible (non-repeating) values.

5. In an AC Circuit, a sinusoidal voltage is given by $v(t) = 20 \, [\cos(10t) + I \sin(10t)]$.

   The current in the circuit is $i(t) = \dfrac{v(t)}{Z}$ with $Z = 20 + j30$. Z is the impedance, a

   measure of the reactances of the many circuit elements to the push of the voltage. The real and imaginary parts of $v(t)$ or $i(t)$ can be plotted using $\text{Re}(v(t))$ and $\text{Im}(v(t))$ or $\text{Re}(i(t))$ and $\text{Im}(i(t))$. Examine the behavior of the system from $t = 0.0$ to 10.0 s in terms of $v(t)$, $i(t)$, their amplitudes and arguments, the ratio of $v(t)$ to $i(t)$ and plots of their real and imaginary parts.

6. With reference to *Beyond the Basics* #5, how does the relation between $v(t)$ and $i(t)$ change if $Z = 20 - j30$? if $Z = 20$?

7. Show that the multiplication and division of the two complex numbers in exponential form $z_1 = |z_1| e^{j \arg(z_1)}$ and $z_2 = |z_2| e^{j \arg(z_2)}$ matches the results of the same operations applied to their rectangular forms.

8. For the complex function $f(z) = \sqrt{z^3 + 1}$, plot the real and imaginary parts of the function over the domain defined by the corner points $z = -10 - 10j$ and $z = +10 + 10j$. Use Mathcad's Surface Plot routine with the points on the complex plane defined as elements of a two-dimensional matrix $z_{i,j}$.

Chapter 8: Calculus of One Variable

One of the enduring high-points of mathematical history is the development of Calculus at the hands and minds of many, but most notably those of Isaac Newton and Gottfried Leibniz.

Calculus, now in a multitude of forms, was developed to solve the problems posed by dynamic systems. If an equation exists which describes the evolution of a system and includes the forces pushing and pulling at it, how does the system evolve? The study of infinitesimals, the notion of the limit, the many rules and regulations of differentiation and its opposite, integration, all serve to increase the level of understanding necessary in the solution of these differential equations.

In this chapter, we will be exploring the limit process using upper and lower bounds, the derivatives of polynomial, trigonometric, exponential and logarithmic functions, numerical, graphical and symbolic forms of differentiation and integration, the interpretation of the integral as an area and a variety of applications.

Remember that the many rules and shortcuts of integration and differentiation are not the heart of calculus. The reason remains the solution of the dynamic relations used to model all systems.

8.1 Evaluating Limits Using Sandwiches

The limit of a function can be viewed as the tendency of the function to move toward a particular value. As the input varies toward a target, it drags the output along and defines the direction in which the output is headed.

The study of the limit is crucial to the fundamental understanding of Newton's (or Leibniz's) Calculus. Calculus is the language of dynamic systems, systems which evolve with time. Its many rules and techniques mean little without applying them to the solution of the questions of Calculus. These differential equations will be explored in further sections.

We will be exploring numerical and graphical techniques useful in determining the limit of a function at a point. In a visual sense, the process is equivalent to zooming in on the area of interest as you would when using a microscope. Numerically, the process of the limit forces a value by tightening the upper and lower bounds placed on the output.

Each increase in the magnification level reveals more and more detail. Classically, our knowledge is then limited by the depth of our perception and the fineness of our measuring instruments.

Warmup

First Case

For the function, $f(x) = \dfrac{x^3}{x+2}$ the behavior of the function as x approaches 2 can be seen directly upon substituting $x = 2$ into the equation. There are no inconsistencies created by this process. The function tends toward $f(2) = 2$. In fact, this value *is* its value at $x = 2$. Figure 1 shows the function's behavior in this area.

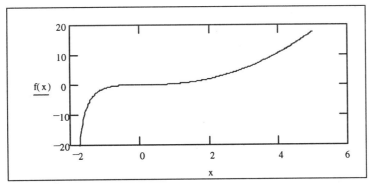

Figure 1

Second Case

However, in the region $x = -2$, problems occur. The denominator is then zero and the function is undefined. Figure 2 shows a large discontinuity in this region. The almost vertical line is an artifact, discussed in Section 3.6, Graphing Singularities and Discontinuities.

Does the function now tend toward a finite value? Certainly, it doesn't in the absolute sense of $f(2) = 2$. But the function does exhibit a smooth behavior toward the discontinuity. As the value of x increases from values more negative than $x = -2$, the function moves toward positive infinity. On the other side of the discontinuity, as the value of x decreases, the function plunges toward negative infinity. This behavior could also be examined by substituting values of x into the function and examining the output as the input values move toward $x = -2$ from each direction.

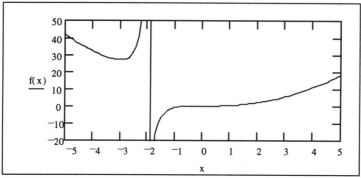

Figure 2

Whereas in the first case, the direction of motion of the input variable toward $x = 2$ would have had no effect on the function's limit, in this case the answers are an infinity apart (or perhaps two infinities).

Third Case

A function expressed as a quotient may have zeros in both the denominator and the numerator when evaluated at a point. A graph may even show the function to be regular and seemingly well-behaved at this point. However, this plot irregularity has more to do with the software you are using than the actual behavior of the function. Some plotters will show the point as a hole in the graph, a place where the function is undefined. Others will smooth right over it. Does a limit exist for these types of functions?

The function $f(x) = \dfrac{x^2 - 4}{x - 2}$ has zeros in the denominator and the numerator for $x = 2$. The function cannot be evaluated at this point. Yet, a plot of the function (Figure 3) in the region of this point shows the function to be relatively well-behaved, finite and smoothly varying. In fact, this twisted rational function looks pretty close to a perfectly straight line. Trying to evaluate the function at $x = 2$ would show that indeed there is a discontinuity there.

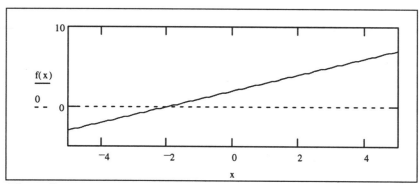

Figure 3

☑ If you attempt to evaluate the function as suggested, you may find that Mathcad returns a value of $f(2) = 0$. What has happened to the expected discontinuity? Mathcad has ordered the operations in such a way that the numerator was evaluated first and found to be zero. Any expression it then multiplied was then zero, right? Wrong! You can see this for yourself by defining the function in the two following ways as shown in Figure 4 and seeing how Mathcad handles the evaluation.

$$\text{a)} \qquad f(x) := \frac{x^2 - 4}{x - 2} \qquad\qquad \text{b)} \qquad f(x) := \frac{1}{x - 2} \cdot \left(x^2 - 4\right)$$

Figure 4

We can examine the behavior of this function in the region of the discontinuity by using our microscope and zooming in on the region around $x = 2$. By allowing values of x to move toward $x = 2$ from below and above the point, the value to which the function tends will be sandwiched out of the process.

In the Polynomial Functions set of exercises, the use of the vector was introduced. We can create a zoom vector by defining a small difference Δ from $x = 2$, evaluating $f(x + \Delta)$ and $f(x - \Delta)$ and then successively reducing the difference by a constant factor. The value toward which $f(x)$ tends then appears (if it is to appear at all). The variable Δ could be defined as the vector:

$$\Delta_K := 0.1^K \text{ with } K := 1 .. 10$$

The values of the function above and below $x = 2$ are evaluated using $f(x + \Delta_K)$ and $f(x - \Delta_K)$ respectively. Figure 5 shows that as the value $x = 2$ is approached, $f(x)$ tends definitely to 4.

$x := 2$ $\quad \Delta := 0.1$	$K := 1 .. 10$ $\quad \Delta_K := 0.1^K$		
$x + \Delta_K$	$f\left(x + \Delta_K\right)$	$x - \Delta_K$	$f\left(x - \Delta_K\right)$
2.1	4.099999999999998	1.9	3.899999999999998
2.01	4.009999999999977	1.99	3.989999999999979
2.001	4.00100000000014	1.999	3.99899999999986
2.0001	4.000099999999391	1.9999	3.999900000000607
2.00001	4.000010000000827	1.99999	3.999989999999173
2.000001	4.000001000088901	1.999999	3.999998999911099
2.0000001	4.000000097699626	1.9999999	3.999999897859482
2.00000001	4	1.99999999	4
2.000000001	4	1.999999999	4
2.0000000001	4	1.9999999999	4

Figure 5

This method is certainly not elegant. Nor is it a proof of the existence of the limit at the point. Definitely not. It is heavy-handed and clumsy but it works well enough to give you an idea of the value of the limit. And it makes you aware of the behavior of the function at the microscopic level.

Explorations

The Basics

1. For the following functions determine $\lim_{x \to a} f(x)$, the limit of $f(x)$ as x approaches a.
   Try using both visual and numeric techniques.

   a) $f(x) = \dfrac{x^4 + 3x^2}{x + 3}$ as $x \to 4$ b) $f(x) = \dfrac{1}{x}\left(\dfrac{1}{8 + x} - \dfrac{1}{8}\right)$ as $x \to 0$

   c) $f(x) = \dfrac{\sin(3x)}{2x}$ as $x \to 0$ d) $f(x) = \dfrac{18x^3 + 9x^2 - 2x}{3x + 2}$ as $x \to \dfrac{3}{2}$

   e) $f(x) = \dfrac{x^4 - 16}{x - 2}$ as $x \to 2$

2. The value to which x moves can be $\pm\infty$. This creates the problem of trying to divide infinitely large numbers by other infinitely large numbers. For each of the following, determine the limit of the function as x approaches infinity.

   a) $f(x) = \dfrac{3x^2 + 5x + 2}{2 - 5x^2}$ as $x \to +\infty$ b) $f(x) = \dfrac{3 - 6x}{x + 2}$ as $x \to -\infty$

3. Even though the Sandwich Technique moves the point x into its target area from both directions, limits can simply be evaluated from one direction. Determine the limit of the function $f(x)$ as x approaches 2 from values less than 2 and from values greater than 2.

   a) $f(x) = \dfrac{x^2 - 5}{x - 2}$ as $x \to 2^-$ b) $f(x) = \dfrac{x^2 - 5}{x - 2}$ as $x \to 2^+$

Beyond the Basics

1. Mathcad's Symbolic Processor can be used to factor any of the expressions in the above functions. For those numerators and denominators which produce zeros in the limit, use this symbolic tool to:
   a) determine the reason the limit can exist at a point although the function may not be defined there. What does the function reduce to everywhere but at the discontinuity?
   b) determine the degree of the plotted polynomial as it appears everywhere but at the discontinuity.

8.2 Derivatives of Polynomial Functions

The odometer of a car gives you a sense of where you are, of how far you have traveled. The speedometer, which measures the rate of change of distance, gives you a sense of movement, of how quickly you will reach your destination.

A mathematical function is dynamic in the same sense. The rate of change of the function at a point gives an indication of the direction the function will take as the input increases. It also defines how quickly the function is changing, whether small changes in the output create large or small, positive or negative output variations.

Given a starting point for the function, a map of the rate of change over the whole domain could be used to rebuild the function itself. The function is contained within its rate of change as the rate of change is contained within the function. This is the basis of the solutions to differential equations. And, this is one of the prime reasons for doing Calculus.

While the odometer readings in the car can be useful for determining the average speed you've been traveling over the duration of the trip, the speedometer indicates the present speed. There is a problem in defining words such as 'instant', 'present' and 'now' since they can be misinterpreted as lengths of time of NO duration!

Practically, the 'now' of the speedometer is a small but finite length of time. The 'now' of a function is controlled by the limit of the change in the input variable. And, as the input changes the output follows it. The slope between two points on the curve varies as one point moves infinitely close to the other. The secant of the two points (the average rate of change) moves toward the tangent (the instantaneous rate of change).

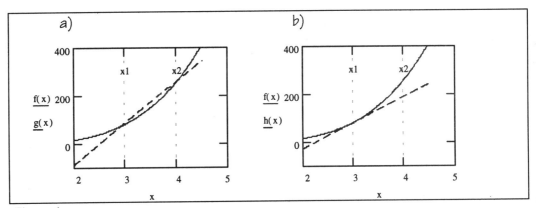

Figure 1

In Figure 1(a) above, $f(x) = x^4$ represents the function while $g(x)$ is the secant line between the points $(3, 81)$ and $(4, 256)$. As the x2 marker moves toward x1 along the *x*-axis and drags the dependent variable value along, the secant line would rotate clockwise about the point $(3, 81)$. Finally, as x2 is infinitely close to x1, the two intersection points merge into the one contact point of the tangent to the curve, $h(x)$. See Figure 1(b).

Warmup

The function $f(x) = 3x^4 + x^2$ in Figure 2 displays the characteristic quartic behavior, a bowl shape opening upwards from an area near the origin. The point (-2, 52) has been marked along the curve.

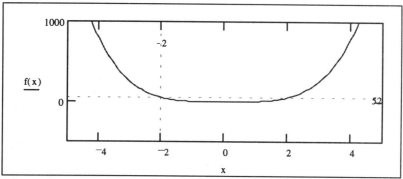

Figure 2

The rate of change of this function at a point can be estimated by moving our point of view closer to the function.

If you fly a plane high above the ground, detail of the surface is obscured. Flight at high altitudes reveals global behavior, overall characteristics which are hidden from a surface-bound observer. However, as you fly closer to the ground, the surface detail of the ground-function reveals itself and the global behavior becomes obscured.

As we move ever inward, using either the Zoom function under the X-Y Plot menu or by simply redefining our plot limits, the curve appears to become linear. See Figure 3. Locally, in small regions about the point of interest, the curve appears to have the characteristics of a straight line. The rate of change or slope of this line can be estimated by examining the ratio of the rise to the run as indicated within the boxed area.

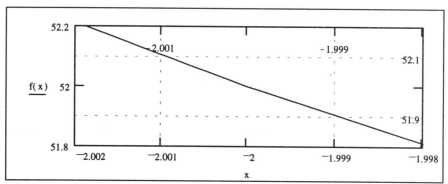

Figure 3

If the box markers in Figure 3 are used to determine the immediate rate of change, the slope of the function at $x = -2$ is then approximately -100.

We can also examine the process of the limit by defining the point of interest x and an increment Δx. By repeatedly reducing the value of Δx, the limit can be estimated. Figure 4 shows the routine for one definition of Δx. The result agrees with our previous estimate.

$xs := 2$ initialize variable to any value $\Delta xs := 0.000001$ delta

$slope(xs) := \dfrac{f(xs + \Delta xs) - f(xs)}{\Delta xs}$ definition of slope in region of point

$slope(-2) = -99.999926995054$ rate of change at x = -2

Figure 4

To repeat this process for every point on the curve may be useful as a form of punishment but is self-defeating. Mathcad offers a variety of more effective ways of examining the derivative at a point.

Mathcad offers a built-in numerical derivative calculator which can be used for individual points or for the function as a whole. By selecting the derivative icon from the vertical palette (v5.0), from the Calculus Palette (V6.0) or by *typing* "?" , a differentiation symbol will appear into which the function and the differentiation variable can be input. The algorithm used works well for points away from discontinuities and singularities and is usually accurate to within the 7th or 8th significant digit.

The analytic or algebraic derivative of the function can be generated by selecting the differentiation variable and, after having loaded the Symbolic Processor, using the Differentiate on Variable option under the Symbolic menu. Or you can simply generate it by hand from the rules of differentiation. Either way provides an exact derivative for the function. This derivative's value can then be compared to the numerical value as shown in Figure 5.

$f(x) := 3 \cdot x^4 + x^2$ ...definition of function

$3 \cdot x^4 + x^2$ by differentiation yields $12 \cdot x^3 + 2 \cdot x$

... use Differentiate on Variable, expression copied to new function dfan(x)

$dfnum(x) := \dfrac{d}{dx} f(x)$ $dfan(x) := 12 \cdot x^3 + 2 \cdot x$ ...definition of numerical

and analytic derivatives

$f(-2) = 52$ $dfnum(-2) = -100$ ...value of function and derivative at x = -2

$dfnum(-2) - dfan(-2) = 1.705 \cdot 10^{-13}$ ...error between numerical and analytic methods

Figure 5

Once defined, the function and its derivative can then be plotted on the same graph so that their relative behavior can be seen. Notice how one function tracks the other in Figure 6. The function and its derivative functions are closely linked.

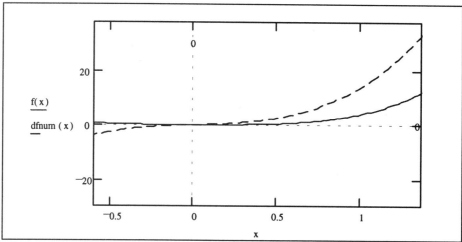

Figure 6

Since first derivatives are themselves functions, they too can have derivatives. The rate of change of the rate of change is the second derivative. Figure 7 shows the function $f(x) = x^4 - 8x^2$ with its first derivative, df1(x), and second derivative, df2(x), overlaid.

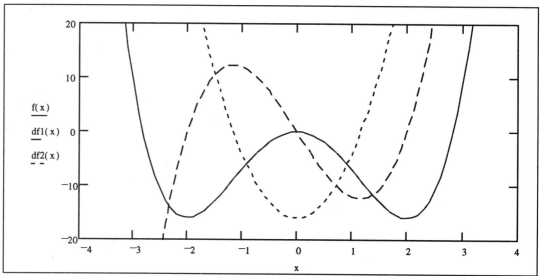

Figure 7

Explorations

The Basics

For the functions given below, examine the original function, its first and second derivatives using numerical, graphical and algebraic techniques. The second derivative is the *first* derivative of the first derivative.

Note how the behavior of one function (i.e., positive or negative outputs, zero outputs, rapid versus slow increases, ...) connect to well defined behavior in the other plots. Try to make your observations and connections explicitly. For example, how does the behavior of the second derivative indicate not only the from of the first derivative but also that of the function itself?

Depending on the plot limits you choose initially, some of the function's and its derivatives' behavior may be obscured. Remember to try flying low to the ground with your observation point close to the function.

a) $f(x) = 6x^2 - 4x + 2$

b) $f(x) = 3x^5 + x^3 - 2x^2 + 7$

c) $f(x) = -x^3 + 4x^2 + 10$

d) $f(x) = \dfrac{6x^2 + 3x + 2}{x^2 - 1}$

Beyond the Basics

1. Develop a file which plots a polynomial function over a given domain and overlays a tangent line at any specified point(s).
2. In the Warmup section, the numerical derivative was compared to the analytic derivative at a point. Extend this idea to include the whole domain. Why does the process seem to exhibit more variation and noise near the endpoints of the domain and little or none near the origin?
3. For function d) from *The Basics*, examine the precision and behavior of the numerical derivative process near a singularity or discontinuity.
4. A crude estimate of the slope at a point was generated using a small Δx routine in the Warmup section. A slope-at-a-point function can be created with Δx as the input variable. Examine the behavior of the function as Δx moves toward zero and compare the final value of that limit process to that generated using the numerical derivative routine.

8.3 The Root Function: Secant and Newton Methods

In an earlier Exploration, the root(s) of a function was (were) determined using a variety of techniques: graphs, Solve Blocks, Solve for Variable and the root (*f(x-guess)*, *x-guess*) function.

At these values of *x*, the function $f(x) = 0$. Graphically, the roots indicate the positions at which the function cuts through the *x*-axis.

To use Mathcad's 'root' function effectively, a plot was made of the function and a visual estimate of the root was determined. This estimate then allowed the iterative process to focus its attention in the immediate vicinity of the estimate rather than searching on its own for a possible root and then finding that the routine did not converge to an answer.

The routine root(*f(xguess)*, *xguess*) returned a value of the root to an accuracy defined by the Tolerance and Precision levels you had specified. The process went on 'beneath the page' as a subroutine prompted by the use of the correct words and syntax. As such, the apparition of the root must have seemed mysterious and magical.

We will develop two ROOT routines using the basic conditional statements (if, until) available in Mathcad. These routines we develop will be highly visible and explicit. Even though Mathcad is not a computer language, it does follow certain structures which allow you to develop an algorithm.

The two methods given here and the third (to be researched) represent different strategies to the solution of the same problem.

Warmup

Secant Method

The Secant Method uses two starting points x_0 and x_1 , one on either side of the root. These points can be obtained from visual estimates. A secant is drawn between these point and its intersection with the *x*-axis is examined. If the process does not return a root at this stage then an iterative process is followed:

I.   Calculate the value of the function at x_0.
II. If $|f(x_0)|$ is less than the tolerance level, the accuracy to which the answer is required, x_0 can be assumed to be the root.
III. Otherwise, calculate $f(x_1)$.
IV. Determine the equation of the straight line containing the points $(x_0, f(x_0))$ and $(x_1, f(x_1))$.
V.   Determine the intersection of this straight line with the *x*-axis.
VI. If the intersection point *x-int* makes $f(x\text{-}int)$ < tolerance then return *x-int* as the root.
VII. Otherwise, use $f(x_1)$ and $f(x\text{-}int)$ as the two new points and repeat the process until convergence is reached.

☑ Even with Mathcad's built-in 'root' function, the attempt to find the root does not always converge. The same applies to Newton's technique. You may have to adjust the number of iterations or the tolerance to achieve the desired effect.

As a test, examine the function $y(x) = x^2 - 4$. By simple analysis, the roots are -2 and +2. A plot of the function (Figure 1) shows that $x_0 = 1$ and $x_1 = 3$ would be suitable choices for the limits of the right-hand-side root.

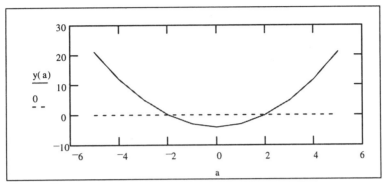

Figure 1

The problem within this and the next method is not so much the algorithm. For a simple curve you can do this procedure using a hand-held calculator with little effort. And, vectors and the use of an index are well-used tools by now. But how do we cause the process to cycle through until the tolerance condition is reached?

Mathcad offers three tools or utilities which can be used in iterative procedures:

1) if(*conditional statement, truevalue, falsevalue*)

This is analogous to the if-then-else statements common to programming languages and has already been introduced in Section 3.6 on plotting discontinuous functions. If the condition is non-zero or true, the *truevalue* (a value or a function) is returned. If the condition is zero or false, the *falsevalue* is returned. This function can be used for scalar equations and can be nested. 'If' statements can exist within the *truevalue* or *falsevalue* of another 'if' statement. Relational operators can also be combined using Boolean algebra to create more complicated conditional statements.

2) until(*condition, indexed equation*)

In an vector equation, the 'until' function cycles the equation with an increase of 1 in the index for every cycle until the condition returns a negative value. At that point, the iteration ceases and Mathcad continues with the document below the region. The equation for the partial sums of a series would be a likely candidate for the application of this function. A likely candidate for the conditional statement would be the error between $f(x\text{-}root)$ and the tolerance.

3) last(vector name)

If you need to count the number of iterations your specific routine took to converge to an answer, use last($x\text{-}root$), where $x\text{-}root$ is the vector name of the successive estimates to the root on its way to convergence. For example, if your process cycles through 10 times to finally meet the tolerance conditions set in your conditional statement, then N : = last($x\text{-}root$) would return a value of 10 for N. This may then be used to display the value of the root.

Figure 2 shows the definition of the iteration limit and the tolerance, the initialization of the x_0 and x_1 points and the definition of the conditional statement for the Secant Method. The tables show the output of the iterative process and reveal a root of +2.0 as predicted from analysis.

$N := 20$ ...maximum number of iterations

$i := 1..N$ ...set iteration and tolerance $T := 10$ $TOL := 10^{-T}$

$x_0 := 1$ $x_1 := 3$ ...initial guesses based on visual estimates

$$x_{i+1} := until\left(\left|y(x_i)\right| - TOL, x_i - y(x_i) \cdot \frac{x_i - x_{i-1}}{y(x_i) - y(x_{i-1})}\right)$$

$index := last(x) - 1$ $j := 1.. index$

j	x_j	$y(x_j)$
1	3.0000000000	5
2	1.7500000000	-0.9375
3	1.9473684211	-0.2077562327
4	2.0035587189	0.0142475399
5	1.9999525932	$-1.8962513479 \cdot 10^{-4}$
6	1.9999999579	$-1.6855966756 \cdot 10^{-7}$
7	2.0000000000	$1.9984014443 \cdot 10^{-12}$

root is $x_{index} = 2.0000000000$

Figure 2

With the tolerance set to 10^{-10}, the root is better defined as 2 accurate to 10 decimal places. Try changing the tolerance to smaller and larger values and examine the limits of the calculation.

With the tolerance set to its default value of 0.001 and a maximum of 20 iterations, the 'until' conditional cycled through 6 times and delivered a root of 2.0000 accurate to 4 decimal places.

High precision is a desirable quality in a number. However, if the 'highly accurate' number generated from your root method is mixed with others of lower accuracy, the computation time necessary to achieve a minimum of error is wasted.

Newton's Method

As this name suggests, Isaac Newton was responsible for the creation of this method. It uses the derivative of the function and one initial guess as a method of zeroing in on the root of the function.

The method uses an estimate to lock the iteration process to a particular region in the plane. You may want to examine the effect, if any, the value of this guess-value has on the number of iterations (and the time taken) the programs needs to converge to a point.

I.   Select a guess value for the root, *x-guess*.
II. Calculate the value of the function at this point, *f(x-guess)*.
III. Calculate the value of the instantaneous slope at this point, $f'(x\text{-}guess)$.
IV. Determine the linear function, $y = mx + b$, having the slope $f'(x\text{-}guess)$ and sharing the common point (*x-guess*, *f(x-guess)*).
V.   Extend this line to the *x* axis and determine its intersection point. This does not require the 'root' function. The root of a linear function is easily solved by rearranging the equation. For the straight line, if $y = 0$ then $x = \frac{-b}{m}$.

VI. If the absolute value of the function evaluated at this intersection point is less than the user-defined tolerance, you have found yourself a root. Otherwise...

VII. Use *x-intercept* as your new guess and repeat the process until the tolerance condition is met and a root is determined.

To begin the process, a guess has to be specified. A plot of the function gives a clear indication of the region to be examined and as a first step, the system defined TOL of 0.001 can be used.

For the function $f(x) = x^3 - x^2 - 1$, Figure 3 shows that an initial guess of $x = 1$ is reasonable for the one existing root.

The maximum number of iterations allowed was 20. And the tolerance was redefined as 0.001, the default value.

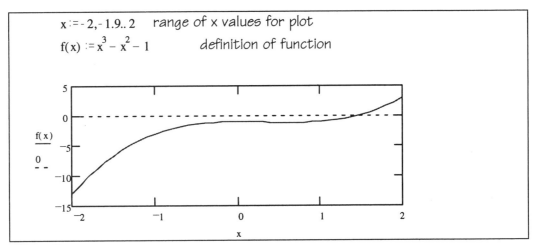

Figure 3

After determining the value of the function, $f(x_0) = 1$, and its derivative, $f'(x_0) = 1$, at the guess point x_0, the iterative process can be launched. The conditional statement and the equation which defines the position of the intersection of the straight line with the x-axis are clearly defined within an 'until' loop for the index $i := 1 .. 20$. This process is illustrated in Figure 4.

$$T := 10 \qquad \text{degree of tolerance} \qquad TOL := 10^{-T} \qquad \text{set tolerance}$$

$$N := 20 \qquad\qquad\qquad \text{max \# of iterations}$$

$$x_0 := 1.0 \qquad\qquad\qquad \text{input initial guess}$$

$$f(x_0) = -1 \qquad\qquad \text{function value at guess}$$

$$df(x) := \frac{d}{dx} f(x) \qquad \text{derivative defined}$$

$$df(x_0) = 1 \qquad\qquad \text{initial derivative value at guess}$$

$$i := 1 .. N \qquad\qquad \text{iteration index}$$

conditional iteration statement determines x intercept until
|f(xint)| is less than tolerance

$$x_i := until\left(\left| f(x_{i-1}) \right| - TOL, \frac{df(x_{i-1}) \cdot x_{i-1} - f(x_{i-1})}{df(x_{i-1})} \right)$$

Figure 4

Although the 'until' expression in Figure 4 may appear initially ominous, a little algebraic manipulation of the expressions in steps 4 and 5 will show this result clearly. The iterative process will continue until the difference between the function evaluated at the last point and the tolerance is less than zero. Then a value of the root is returned by defining the root as the last entry in the x vector.

For the default tolerance, the routine went through 6 cycles to reach a value of $x = 1.4656$. Since the x-index started from 0 (the initial point) the index output by the last(x) function had to be reduced by 1.

With the tolerance set to 10^{-10}, the routine cycled through only one more time and output a value of $x = 1.4655712319$. The output is shown in Figure 5. At this point, you may ask yourself how well you need to know this root value. Good question!

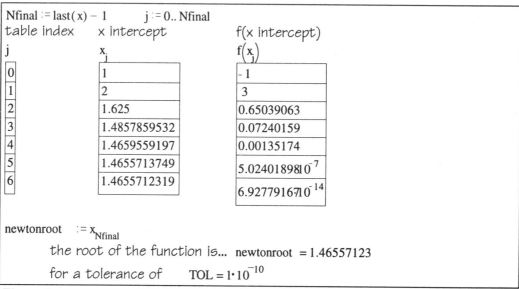

Figure 5

Explorations

The Basics

Use the two techniques discussed in the Warmup to determine all the roots of the following polynomials.

1. $y(x) = 2x^3 - 6x^2 + 5x - 1$

2. $y(x) = 6x^2 - 10x - 4$

3. $y(x) = x^4 + 2x^3 - 4x^2 + 1$

4. $y(x) = e^{-x} \sin(x)$ for the domain $x = +2$ to $x = +11$

Adjust the tolerance settings within each file and examine any increases in computation time. Compare the methods for clarity, precision, and ease of use.

Beyond the Basics

1. Research and program the method of midpoints, a simple yet effective tool for determining the roots of an equation. Compare and evaluate the method versus the other two developed so far.

As the name suggests, this iterative procedure examines the value of the function $f(x)$ at a midpoint between two user-specified values of x which sandwich the root. If certain conditions are met, the value of $f(x\text{-}midpoint)$ approaches zero and $x\text{-}midpoint$ approaches $x\text{-}root$.

The algorithm for the midpoint method can be broken down as follows:

 I. By a visual examination of the function in a certain region (where $f(x)$ is close to zero), determine two values of x, x_1 and x_2, one on each side of the suspected root.

 II. Calculate the average value of the x coordinates. The average value is defined as $x\text{-}average = (x_1 + x_2)/2$.

 III. Calculate the value of $f(x\text{-}midpoint)$.

 IV. If the absolute value of $f(x\text{-}midpoint)$ is less than the tolerance you've set, you've found the root. If not, then...

 V. Calculate $f(x_1)$ and $f(x\text{-}midpoint)$ and determine if $x\text{-}midpoint$ is to the left or to the right of the root by comparing the signs (+ or -) of $f(x_1)$ and $f(x\text{-}midpoint)$.

 VI. If $f(x_1)$ and $f(x\text{-}midpoint)$ have opposite signs, use $x\text{-}midpoint$ and x_1 as the pair of x values which sandwich the root and repeat the process until the absolute value of $f(x\text{-}midpoint)$ is below your tolerance setting.

Or...

 VI. If $f(x_1)$ and $f(x\text{-}midpoint)$ have the same signs (they can't be sandwiching the root) then replace x_1 with x_2. The value pair, $f(x\text{-}midpoint)$ and $f(x_2)$ should now have opposite signs.

 VII. Use $x\text{-}midpoint$ and x_2 as the new pair of x values which sandwich the root and repeat the process until the absolute value of $f(x\text{-}midpoint)$ is below your tolerance setting.

2. Research and program the method of Picard (simple iteration) and compare and evaluate the efficiency of the three iterative methods (Secants, Newton, Picard) as applied to one function. This method first substitutes an estimate into the equation and searches for consistency (i.e., the x-value substituted into one side of the equation should generate the same x-value on the other side of the equation). The process is repeated until a consistent answer is determined to the required tolerance.

8.4 Critical Points in Polynomial Functions

In Section 8.2, Derivatives of Polynomial Functions, we explored the behavior of a function and its rate of change. In one part of *The Basics* exercise, the relationships between the function and its derivatives were examined. The behavior of the function is mirrored in the behavior of its derivatives. There is a tight web between them.

Here, we examine the 'critical points' of the function, points on the curve which have distinctive features and qualify the behavior of the function. Knowledge of the location of these points alone allows a quick if possibly inaccurate visualization of the function.

In a system, these points represent areas of stability, of maximum or minimum output, of changes of sign in the rate of change of the function and of changes in the curvature of the function.

And, keep in mind, whatever shortcuts for differentiation there are, the purpose of Calculus is to solve differential equations.

Warmup

The polynomials we have explored all seem to exhibit the same behavior: a region of curves, hills and valleys, followed by a more linear move toward positive or negative infinity in the wings of the domain.

The First Derivative

Over a finite domain, the curve $f(x) = 8x^4 - 7x^3 - 2x^2 + x + 2$ in Figure 1 shows the characteristic polynomial behavior.

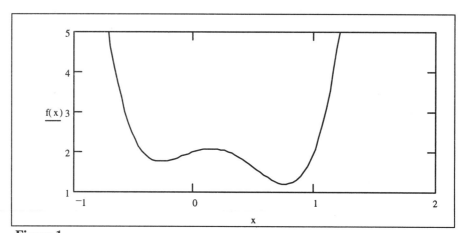

Figure 1

Over the domain $x = [-1,2]$, the plot shows three points where the direction of the function changes. The first (from the left) is near $x_1 = -0.3$. Then the function changes again near $x_2 = 0.2$. And, finally, near $x_3 = 0.8$.

Tables of values for the function near each of these points, given in Figure 2, show where the turning points occur. Further numerical or graphical zooming would then reveal the exact inputs, x_1 and x_3 , which produce the minimum output in their respective regions and the exact input x_2 which produces a maximum output. On the restricted domain $x = [-1,2]$, these (x,y) coordinates represent *local* points. The end points are the maximum values the function has over its range. In that respect, x_3 also indicates a *global* minimum.

x_1	$f(x_1)$		x_2	$f(x_2)$		x_3	$f(x_3)$
-0.4	1.9328		0	2		0.6	1.4048
-0.35	1.825175		0.05	2.044175		0.65	1.310675
-0.3	1.7738		0.1	2.0738		0.7	1.2398
-0.25	1.765625		0.15	2.085425		0.75	1.203125
-0.2	1.7888		0.2	2.0768		0.8	1.2128
-0.15	1.832675		0.25	2.046875		0.85	1.282175
-0.1	1.8878		0.3	1.9958		0.9	1.4258

Figure 2

When we examine tables of values for the function and its derivative in the region of a minimum, we note that the rate of change (the first derivative) seems to be going through a transition itself. For the x_1 point region, the table of values, Figure 3, shows the first derivative changing sign from negative to positive as the local minimum is passed. To change its sign, the derivative must pass through zero. This then could be the indicator for a minimum (or a maximum).

x_1	$f(x_1)$	$\dfrac{d}{dx_1}f(x_1)$
-0.4	1.9328	
-0.35	1.825175	-2.8080
-0.3	1.7738	-1.5445
-0.25	1.765625	-0.5540
-0.2	1.7888	0.1875
-0.15	1.832675	0.7040
-0.1	1.8878	1.0195
		1.1580

Figure 3

There is then a correspondence between the change of sign of the derivative, its passage through zero, and the existence of a critical point. However, you have to be careful here. Although a minimum or maximum requires a slope of zero at the point, a slope of zero does not necessarily mean there is a minimum or a maximum.

It seems then that a way to determine these points is to examine the behavior of the first derivative. By overlaying the function and its derivative, we can see that the min/max points correspond to those values of x for which the first derivative is zero.

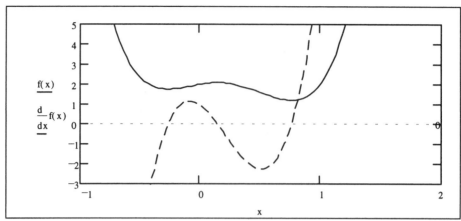

Figure 4

The first derivative function's zeros can be determined using any of the techniques developed for calculating the roots of a function. With an estimate of *xguess* = -0.4, the root function applied to the derivative returns:

derivative function	$df(x) := \frac{d}{dx} f(x)$
the guess	$xguess := -0.4$
root definition	$x_1 := root(df(xguess), xguess)$
root output	$x_1 = -0.264305$
minimum value of	$f(x_1) = 1.764266$

Figure 5

The first derivative exhibits the same type of polynomial behavior as the function. As it should since, analytically, its functional relationship is $f'(x) = 32x^3 - 21x^2 - 4x + 1$ from the application of the Power Rule. Its minimum and maximum values can then be pinpointed by the zeros of its derivative, the second derivative of the function. What do these points tell us about the behavior of the $f(x)$?

The Second Derivative

Figure 6 shows the function $f(x)$, its first derivative $d_1(x)$ and the second derivative $d_2(x)$. As in the case of the function, the peaks and valleys of the first derivative are characterized by a second derivative of zero. The location of this zero corresponds to a point on the function where the curvature is changing, where the function is moving from a concave upward to a concave downward behavior or vice versa.

The first derivative is increasing positive from the left. At the first point where $f''(x) = 0$, the first derivative switches directions and becomes decreasing positive. The function mirrors this by changing its concavity from upward to downward at this *point of inflection*. How would you describe the behavior of the function at the minimum point for $f'(x)$?

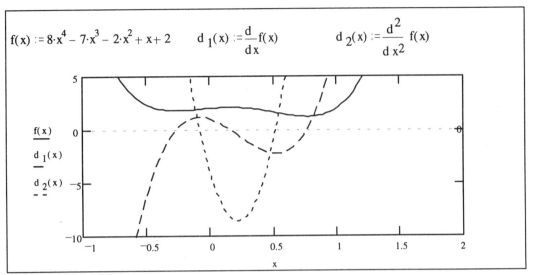

$$f(x) := 8 \cdot x^4 - 7 \cdot x^3 - 2 \cdot x^2 + x + 2 \qquad d_1(x) := \frac{d}{dx} f(x) \qquad d_2(x) := \frac{d^2}{dx^2} f(x)$$

Figure 6

☑ The second derivative can either be generated by defining another function as the derivative of the first derivative function or by using the higher derivative definition in Mathcad. This function is created by selecting from the Calculus (v6.0) or symbol (v5.0) palettes or by typing CTRL - ? and filling in the appropriate place-holders.

This process could go on until there is nothing left of the *n*th order derivative. However, an analysis up to the second order provides us with more than enough information to characterize a function's behavior. For higher order derivatives, the process simply cycles through the same routine.

Again, care must be exercised in applying these connections. A $f''(x) = 0$ does not necessarily indicate a minimum or maximum in the first derivative.

Explorations

The Basics

1. Construct a cubic function which has $f'(0) = 0$ yet exhibits neither a maximum nor a minimum at this point. Does this behavior apply to other polynomials of higher degree?

2. Analyze each of the following functions and determine the minimum and maximum points and any points of inflection:
   a) $y(x) = x^2 - 3x$
   b) $y(x) = x^3 - 3x^2 = 24$
   c) $y(x) = 4x^4 - 3x^3$
   d) $y(x) = (x - 2)^3 (x + 1)^2$

3. The function $f(x) = x^5 - 2x^4 + x^3 - 3x^2 + 1$ is much like the one examined in the Warmup section. Make a full analysis of the behavior of this function over the domain [-5, 5]. Try to use as many tools (tables, plots, root function, Solve Blocks, numerical derivatives, symbolic derivatives, ...) as you can to make the behavior explicit and visible.

4. Design an optimum metal-use cylindrical container. For a volume of 341 mL (cm³), the volume of a pop can, determine the radius and height of the can which would minimize the use of sheet metal in its construction. Assume the hollow can forms a perfect closed cylinder. Once this has been solved, apply the technique to other hollow forms: cubes, cones, pyramids,...

5. In Electrical theory, maximum power is transferred from a source to a load when the load resistance matches the internal source resistance. Power is given by

$$P = \frac{V^2}{(R+r)^2} \cdot R$$ where V is the source voltage, R is the load resistance and r is the

fixed internal resistance of the battery. Show that 'resistance matching' holds for the case of a supply of 10 Volts DC with an internal resistance of $r = 20$ Ohms across a variable load R.

Beyond the Basics

1. From the following data set, reconstruct the shape of the original function. The function can be rebuilt to a good approximation as a set of vector points using the spline interpolation tools (lspline, pspline, cspline) available in Mathcad (in Help under Spline Functions).

data point	(-1.00, 17.00)
maximum	(0, -4.00)
point of inflection	(1.11, -12.23)
minimum	(2.22, -20.46)
data point	(4.00, 28.00)

What is the approximate value of the function at:
   a) $x = -3.00$?
   b) $x = 3.25$?
   c) $x = 6.00$?

2. A function has a first derivative given by $g'(x) = 3x^2 - 2x$ and displays a maximum point at (0,2). Create an approximate plot of the original function by predicting the location of the next point from the maximum along the curve. The first derivative gives the rate of change of the function in the immediate vicinity of a point. This next point is then separated from the known point (x_0, y_0) by a small horizontal distance Δx and a small vertical distance $\Delta y = \dfrac{dy(x_0)}{dx} \cdot \Delta x$. Repeat the process until the curve is plotted over a suitable domain and range.

8.5 Derivatives of Trigonometric Functions

The trigonometric functions display a unique kind of self-reference. The standard cosine function is a phase-shifted sine function. The tangent is the ratio of the sine to the cosine. When the rates of change of these functions are examined, the self-reference continues. This quality makes the sine function show up, in one form or another, in all variety of applications and differential equations

As the sine function is the source of all other trigonometric functions, it is the source of the layers and layers of superimposed functions used in the synthesis and analysis of complex waveforms (audio waves or electromagnetic waves). And, the character of its derivatives allows for quick and easy estimation routines where its direct evaluation would be impossible (as in scientific calculators and any computer program using trigonometric functions).

Warmup

The sine function should be a familiar friend by now. The most general form $y = A \sin(Bx + C)$ where x is in radians includes an amplitude A, a frequency index B measured in cycles/2π and a phase angle C related to the phase shift φ by $\varphi = \dfrac{-C}{B}$.

The analytic process of determining the derivative of the sine function is discussed in detail in many texts. Although the process may seem overwhelming at first glance, further analysis shows it to be made up of the limit process and a few trigonometric identities.

As well, the derivatives of the trigonometric functions are usually clearly tabulated for reference. However, can you SEE these definitions?

If we use the Symbolic Processor, any trigonometric function's symbolic derivative can be generated in a fraction of a second. Figure 1 holds a few sine function variations with their derivatives alongside.

$\sin(x)$	by differentiation, yields	$\cos(x)$
$\sin(2 \cdot x)$	by differentiation, yields	$2 \cdot \cos(2 \cdot x)$
$4 \cdot \sin(3 \cdot x + 4)$	by differentiation, yields	$12 \cdot \cos(3 \cdot x + 4)$

Figure 1

All the differentiation rules or shortcuts still apply: power, quotient, product, chain rules. Whereas in the derivative process applied to polynomial, the degree of the function is reduced by one, here the derivative seems to simply produce yet another sine-type function, the cosine.

Let's determine the derivative of the sine function at a few of its critical points and zeros. Figure 2 shows the sine function $y = \sin(x)$ over the domain $[-3\pi, +3\pi]$.

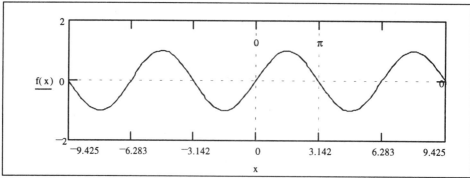

Figure 2

In Figure 3, with the derivative defined as derA(x1), the plot shows that the maximums or minimums have a slope of zero at the critical point. As well, the calculated slopes at the zeros of the function end up as of +1 or -1. The derivative function seems to follow the sine curve very closely. At this stage, the derivative was evaluated every (π/2) radians and the shape seems roughly triangular.

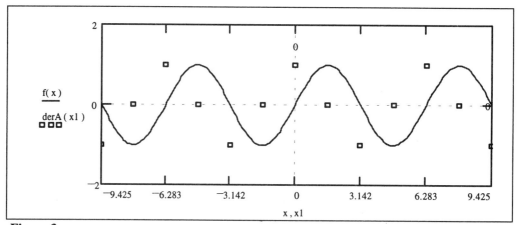

Figure 3

If we now increase the number of evaluated points over the range, as derB(x2) of Figure 4, we start to see a familiar pattern emerge, that of the cosine function.

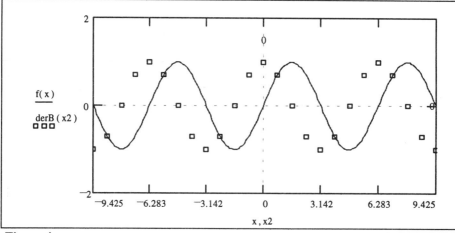

Figure 4

Further increase in the resolution of the derivative as in df(x), Figure 5, shows an exact correlation with the cosine function.

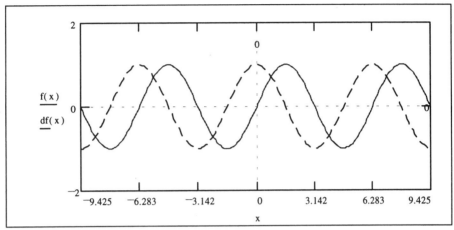

Figure 5

Thus the limit process, whether applied symbolically or numerically, returns a sine function shifted to the left by 90°.

Repeated application of the derivative to a polynomial function will eventually reduce the expression to zero. Here, in the case of the trigonometric functions, the derivative re-expresses the function in familiar terms.

If we apply the second derivative to the sine function, as in ddf(x) of Figure 6, the operation produces almost an exact copy of the original function except for the negative sign. The second derivative of $y = \sin(x)$ (which is the *first* derivative of the cosine function) is $y'' = - \sin(x)$.

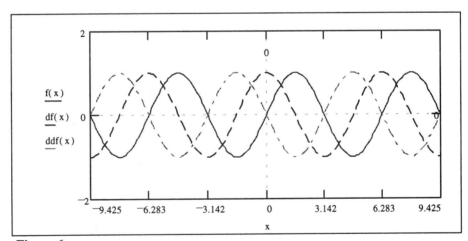

Figure 6

What is the order of the derivative that would copy the sine function onto itself? Is this derivative order unique?

If the sine function were a song, repeated application of the derivative would eventually "bring us back to DO".

Explorations

The Basics

1. Apply the method of analysis developed in the Warmup to the examination of the general sine function $y = A \sin(Bx + C)$. How do the rules of differentiation express themselves in the plots you have created?

2. Examine the rate of change of the function $y(x) = 3\sin(x) + \sin(3x)$ relative to that of its components $y_1(x) = 3\sin(x)$ and $y_2(x) = \sin(3x)$.

3. The tangent function can be expressed as the ratio of the sine function to the cosine function. Show that the symbolic derivative of the tangent is equivalent to the form of the plotted numerical derivative. Examine tangent functions which include $B > 1$ or $B < 1$ and non-zero phase angles.

4. For the reciprocal trigonometric functions, secant, cosecant and cotangent, show that the numerical forms of the plotted derivatives match the symbolically generated derivatives.

Beyond the Basics

1. Show, using plots and analysis, that the function $y_2(x) = -kA \sin(kx)$ is the first derivative of the function $y_1(x) = A \cos(kx)$ for your choice of the constants k and A.

2. For a mass on a spring, the restoring force is given by $F = - kx$ where x is the extension of the spring from its equilibrium position and k is the force constant. Since unbalanced forces cause accelerations, the expression F can be replaced by $m\dfrac{d^2x}{dt^2}$, the mass times the acceleration. The solution of this differential equation $m\dfrac{d^2x}{dt^2} = -kx$ is the function $x(t)$ which satisfies the conditions of the derivative and the initial condition. Show that a possible general solution is $x(t) = x_0 \sin(\omega t)$ with $\omega = \sqrt{\dfrac{k}{m}}$. Show that the solution applies to, in particular, a maximum displacement from equilibrium x_0 of 0.040 m, a mass, m, of 0.010 kg and a spring constant $k = 0.1$ N·m^{-1}. How would the solution characteristics change (if at all) if the sine function were replaced with a cosine function?

8.6 Derivatives of Exponential and Logarithmic Functions

The exponential function deserves a special place in a section on Calculus. It appears repeatedly in the solutions to linear differential equations due to its special properties under differentiation.

Just as the derivatives of sine and cosine functions stay within the family of trigonometric functions, the derivatives of exponential functions remain within their same family. There is a type of symmetry involved within these functions which allows them to retain their form and identity under the transformation of the derivative process.

Unlike the derivatives of polynomials where the repeated application of the derivative ultimately leads to an output of zero, the derivatives of trigonometric and exponential functions seem to lead to yet more trigonometric and exponential functions respectively.

The logarithmic function (i.e., the inverse of the exponential function) behaves in a like manner, generating a succession of inverse powers of x upon repeated differentiation.

Warmup

The exponential function is defined as $y = b^x$ where b is any non-zero and positive number.

Let's examine the exponential function $y = 2^x$ along with its symbolic and numerical derivatives.

After loading the Symbolic Processor and selecting output preferences, we apply the derivative to the expression 2^x. The result is given below in Figure 1(a). If we then take the derivative of this new expression, the same form of expression emerges.

a) 2^x by differentiation, yields $2^x \cdot \ln(2)$ which by differentiation, yields $2^x \cdot \ln(2)^2$

b) $\sin(x)$ by differentiation, yields $\cos(x)$ which by differentiation, yields $-\sin(x)$

Figure 1

The derivative of an exponential function seems to be given by the function itself times a constant $\ln(b)$ generated for each time the derivative is taken. Notice in Figure 1(b), the symbolic derivatives for the sine function end up generating a shifted sine functions (the cosine and negative sine).

What would the expression be for the nth derivative of $y = 2^x$?

The pattern emerges. The effect of the derivative operator on the exponential function is to recreate the original function to within a multiplicative constant. The function is pretty well copied onto itself.

If we apply the numerical derivative to the function and plot both the function and its derivative on one graph, the shape of the derivative confirms the result of the symbolic process.

Figure 2 shows that the derivative creates another exponential function and, since $\ln(2) = 0.693$, this derivative function lies below the $y = 2^x$ line.

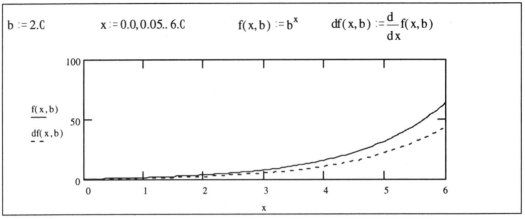

Figure 2

☑ The functions f(x, b) and df(x, b) in Figure 2 have been made functions of two variables to allow for quick easy changes to the base of the exponential function.

Try this same process for any other base, $b = 3$ or 5, applied to $y = b^x$.

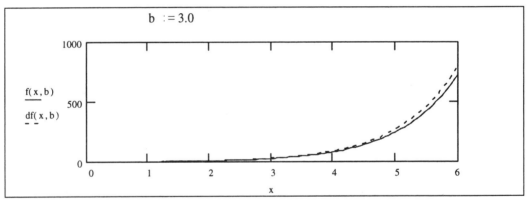

Figure 3

Notice that while the $b = 2$ plot's derivative lies below the function, the derivative of the plot for $b = 3$ (Figure 3) lies above the function line.

Is there a base for which $f(x) = f'(x)$?

If the bases $b = 2$ and $b = 3$ create derivative curves above and below their respective functions, there should be a base between these two for which the derivative maps itself directly back onto the original function. Try to find the value of this particular base by repeated exploration.

To see how well you have established this base, examine the difference between the function and its derivative by plotting the function $differ(x,b) = f(x,b) - df(x,b)$ where $df(x,b)$ is the numerical derivative function.

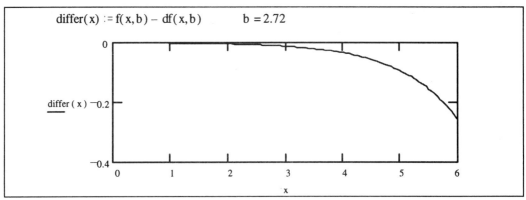

Figure 4

Figure 4 shows the function *differ(x,b)* for a base $b = 2.72$.

The end result of all this directed trial and error should be a reasonably good approximation to the base of the natural logarithm ($e = 2.718...$).

Functions of base e then provide solutions to systems for which the derivative depends on the size of the system, for which growth or decay vary directly with the size of the population.

The derivatives of logarithmic functions will be examined within the Explorations.

Explorations

The Basics

1. Show, using symbolic, graphical and numerical analysis that...
   a) $y(x) = 6e^{-3x}$ has a first derivative of $-18e^{-3x}$
   b) $y(x) = 2e^{-x}$ has a second derivative of $2e^{-x}$
   c) $y(x) = e^{x^2}$ has a first derivative of $e^{x^2} \cdot 2x$
   d) $y(x) = \ln(e^x + 1)$ has a first derivative of $\dfrac{e^x}{e^x + 1}$

2. Determine the rate of change of the following functions and show the graphical relation between the function and its derivative.
   a) $y = 5t^3 e^{2t}$ \qquad\qquad b) $y = 2e^{2x} \sin(x)\cdot$
   c) $y = \sin(2x) \cdot e^{-x^2}$ \qquad d) $y = \ln(\cos x)$

   In d), the cosine function exhibits zeros at which the natural logarithm (ln) is undefined. How do you prevent this problem?

3. A system's behavior in time is controlled by the equation $\dfrac{dy}{dx} - 2y = 0$. Show using plots, tables, numerical and symbolic derivatives that the function $y = 3\,e^{2x}$ is the solution to the differential equation with an initial condition of $x = 0$ and $y = 3$.

4. The derivative of $y = \log(x)$ can be generated using the Symbolic Processor. Examine the reasonableness of this result by using plots and tables of the function and its numerical derivative. What is the general form of the derivative for $y = a \log (kx)$?

5. Repeat the process of #2 for the function $y = \ln(x)$. What is the general form of the derivative for $y = a \ln (kx)$?

6. Determine the point x at which $\log(x) = \dfrac{d \log(x)}{dx}$. You may find that the Symbolic menu option 'Solve for Variable' is unable to return a value for x. What other solution techniques are available?

Beyond the Basics

1. In a series RL (a resistor in series with an inductor) circuit with an applied DC voltage at $t = 0.0$ s, the current as a function of time is given by $i(t) = \dfrac{E}{R}(1 - e^{\frac{-Rt}{L}})$. This is the solution to the linear differential equation describing the dynamic behavior of the circuit at time $t > 0$ seconds. Show, using plots and tables of the function $i(t)$ and its derivative for $t > 0.0$ seconds, that the expression for $i(t)$ satisfies the equation $L\dfrac{di}{dt} + Ri = E$ for your choices of R, L and E. The expression Ri is the voltage across the resistor at a time t while the voltage across the inductor is given by $L\dfrac{di}{dt}$. The differential equation is simply an application of Kirchoff's Law.

2. Euler's Formula $e^{j\theta} = \cos(\theta) + j\sin(\theta)$ will be examined in a later section on the Taylor Series. The output of $\mathrm{Re}[e^{j\theta}]$ and $\mathrm{Im}[e^{j\theta}]$ would be represented by $\cos(\theta)$ and $\sin(\theta)$ respectively. Show using plots of the Real and Imaginary parts of the function and its derivative over the range $\theta = 0$ to 2π that the derivative of the function $e^{j\theta}$ is given by $je^{j\theta}$.

8.7 Definite Integrals and the Area Beneath the Curve

The definition of the Definite Integral is at the heart of Calculus. The Fundamental Theorem of Calculus connects the ideas of the derivative, the anti-derivative, limits of integration and the interpretation of the definite integral as an area under the curve.

In this exercise, we will explore the use of Mathcad's numerical integration process.

Mathcad's process of integration uses a routine which breaks the area defined by the continuous function, the x-axis and the upper and lower limits into a series of simple trapezoids. The summation of these separate areas leads to an estimate of the value of the integral between the upper and lower limits. The process is iterative and, as we have seen with other such processes, depends heavily on the value of the Tolerance defined by the user. Again, we must be aware of just how much accuracy we need.

For a range of $x := 0.00, 0.01 .. 100$, Mathcad's integration routine would have to calculate 10,000 separate integrals, one for each upper limit relative to the lower limit of zero. Within each of those integrations is an iterative routine. The time taken can be considerable depending on the function and the hardware even at average tolerance levels. If you demand a tolerance level defined by $TOL = 10^{-15}$ for a complicated exponentially varying sinusoidal-logarithmic function then expect to be able to go for a short nap while your computer is crunching through its task. Mathcad may just give up if the number of iterations goes beyond its computational limit.

Warmup

Definite and Indefinite Integrals

The derivative of a function is itself a function, a point by point measure of the rate of change. And, as well, the indefinite integral is a function, one whose derivative is the integrand (an antiderivative). The integration constant is specified by the initial conditions imposed on the function.

Mathcad's numerical integration routine cannot return an indefinite integral. That feature is left for the Symbolic Processor and will be examined in the next set of Explorations along with the details of numerical integration.

The *definite* integral is strictly a *single number*. Analytically, it is the result of the difference between the antiderivative $F(x)$ evaluated at the upper and lower range limits.

$$\int_a^b f(x)dx = F(b) - F(a)$$

Numerically, the definite integral is the result of a numerical approximation routine involving summation. However, the process of the definite integral can be molded into resembling an indefinite integral function of sorts. The integral then generates a function point by point over the range of the independent variable.

For the variable *x* over a range with a given step size, plots and tables can be created for a function *f(x)*. The definite integral can likewise be generalized into a function by fixing its lower limit and defining its upper limit in terms of the variable *x*. For the function *f(x)* with *F(x)* as the antiderivative and *F(0)* as the initial value,

$$F(x) - F(0) = \int_0^x f(t)dt$$

Mathcad then evaluates *F(x)* for each of the points *x* as defined in the statement of the domain and returns a value to the numerical precision and tolerance specified. The lower limit, *F(0)*, creates an offset which can be compensated for if required.

☑ Leaving the calculation mode in Manual (Automatic unselected) will probably save you a few headaches in your Explorations.

For the function $f(x) = x^3$ we expect the typical cubic form. By analysis, we know the indefinite integral of a cubic is a quartic (x^4). If we define a definite integral with a varying upper limit as was suggested then we can create an indefinite integral function accurate to within the initial condition specified by the integration constant. Figure 1 includes a plot of the function *f(x)*, its indefinite integral *intf(x)* and a table of values of *intf(x)* which corresponds to the definite integral output as the upper limit varies.

The integration symbol is generated by either selecting the icon from the vertical palette (v5.0) or the Calculus palette (v6.0) or by *typing* '&'.

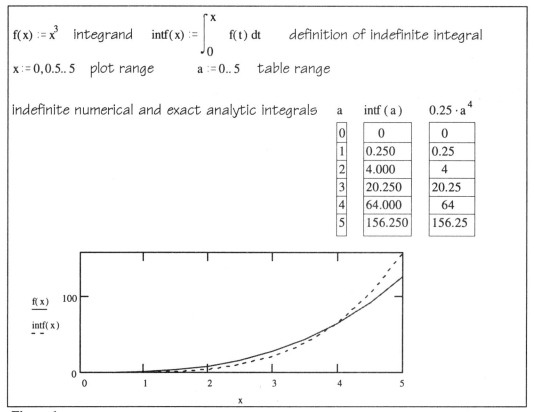

Figure 1

Even though the functional form of the indefinite integral is not specified by the numerical process, we know that the antiderivative of x^3 is $\dfrac{x^4}{4} + C$. With $C = 0$, a comparison of the exact antiderivative function versus the table of values agrees.

The Area Under the Curve

The definite integral can be interpreted as the area under the curve bounded by the function, the x-axis and the integration limits. By dividing the domain into a finite number of intervals and examining the approximate area under the curve for each interval, an estimate of the total area can be determined. If the number of intervals is allowed to grow to infinity, the estimate moves toward an exact representation of the area.

The area under the curve does not necessarily have to be defined from the x-axis . Any suitable reference axis or curve can be used. The height from the lower to the upper curve is the factor which defines the length of the initial approximation rectangles or trapezoids within the summation process.

In fact, the integral defined as the area from the curve to the x-axis is simply one specific case of this more general view. For two curves f(x) and g(x), with f(x) superior to g(x) over the integration, the difference would be defined as f(x) - g(x).

For the x-axis case, g(x) = 0 and is left invisible and (hopefully) understood.

The area between curves f(x) and g(x) then is given by $\displaystyle\int_a^b [f(x) - g(x)]dx$ for $f(x) \geq g(x)$ and with limits x = a to x = b.

If we examined the area defined by the two straight lines $y_1(x) = 2x - 5$ and $y_2(x) = -3x + 6$ between x = -3 and their intersection point, a plot would help us to determine not only which function is superior over the integration range but also the upper limit of that range. The intersection point, where $y_1(x) - y_2(x) = 0$, could be determined using 'Solve Blocks' or the 'Solve for Variable' option under the Symbolic menu.

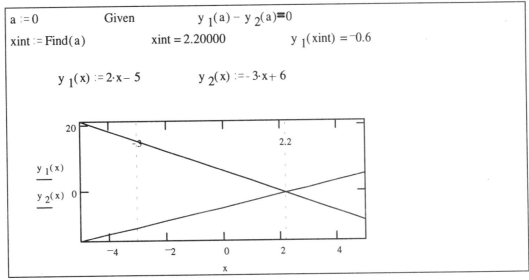

Figure 2

From Figure 2, the intersection occurs at $x = 2.2$. The area between the curves and the integration limits can then be defined as $\displaystyle\int_{-3}^{2.2} (-3x + 6) - (2x - 5)dx$.

The x-axis boundary condition defines any area underneath the x-axis as being negative. The integral of $f(x) = x^3$ from $x = -2$ to $x = +2$ is zero since the areas (negative from $x = -2$ to 0 and positive from $x = 0$ to $+2$) cancel. Although the integral of $y_1(x)$ by itself would create a negative area (-30.16), the subtraction of this area ensures that the total area above $y_1(x)$ and bounded by $y_2(x)$ is measured. The summation process now uses the lower function as its base and measures all heights relative to it.

This same process can also be viewed as the difference between two x-axis reference line integrals. Figure 3 includes the problem expressed as a single integral and as the difference between separate integrals.

$$\int_{-3}^{2.20} y_2(x) - y_1(x)\, dx = 67.6 \qquad\qquad \text{single integrand}$$

$$\int_{-3}^{2.20} y_2(x)\, dx = 37.44 \qquad\qquad \int_{-3}^{2.20} y_1(x)\, dx = -30.16 \quad \text{separate integrals}$$

$$\int_{-3}^{2.20} y_2(x)\, dx - \int_{-3}^{2.20} y_1(x)\, dx = 67.6 \quad \text{difference of separate integrals}$$

Figure 3

Explorations

The Basics

1. Examine the integral of the cubic function $f(x) = 4x^3$ over a symmetric range (e.g.: over the domain [-4, +4]). What do you observe about the results of integrating odd functions over this type of limit. Repeat the process over a non-symmetric range. How could you redefine the integrals to include only the part which does not cancel?
2. Examine the integral of the quadratic function $f(x) = 3x^2 + 4$ over a symmetric range. What do you observe about the results of integrating even functions over this type of range. Repeat the process over a non-symmetric range. Can the symmetric part within the non-symmetric range be simplified? Is the simplification worth the effort?
3. Determine the area enclosed by the functions $y_1(x) = -4x^2 + x + 3$ and $y_2(x) = 6x - 4$. Although there will be areas between the functions in the regions as $|x| \gg 0$, these areas are unbounded.
4. Determine the area between the two quadratic functions:
   $$y_1(x) = -2x^2 + 6x + 3 \quad \text{and} \quad y_2(x) = 4x^2 + 6x - 15.$$
5. Determine the area between the two cubic functions:
   $$y_1(x) = -x^3 + 2x^2 + 6 \quad \text{and} \quad y_2(x) = 2x^3 - 4x^2 + 8.$$
6. Calculate the area enclosed by the circle $x^2 + y^2 = 25$ and the quadratic function $y(x) = x^2 + 4x + 6$.

Beyond the Basics

1. For any of the previously examined functions, examine the effects of Tolerance (TOL) on the calculation of the definite integral value or indefinite integral function. For a specific setting of the Displayed Precision, does the Tolerance affect the result of the integration process. Is there a noticeable increase in computation time for successively smaller Tolerance settings? Do either the time or accuracy effect depend on the relative complexity of the function?

2. Two ellipses sit one inside the other. Determine the cross-sectional area between the ellipses if the outer is defined by $\dfrac{x^2}{36} + \dfrac{y^2}{16} = 1$ while the inner ellipse is defined by $\dfrac{x^2}{4} + \dfrac{y^2}{9} = 1$.

3. Two offset ellipses overlap in an area to the right of the origin. Determine the size of the common region for the ellipses $\dfrac{x^2}{16} + \dfrac{y^2}{25} = 1$ and $\dfrac{(x-6)^2}{25} + \dfrac{y^2}{16} = 1$.

4. A two-dimensional area is bounded by three functions, two of them linear and the third quadratic. The functions are:
$$y_1(x) = 12x + 6, \quad y_2(x) = -10x + 4, \quad \text{and} \quad y_3(x) = -x^2 + 4x + 5.$$
Calculate the area enclosed within the function boundaries with special attention to the region where all three functions intersect.

8.8 Numerical and Symbolic Integration Techniques

Mathcad's integration routine, used in the previous set of Explorations (8.7) is fairly sturdy. The only time you may experience difficulty is in the integration of rapidly varying sinusoidal functions over many cycles.

A symbolic process may seem the answer. Under the 'Symbolic' menu, the 'Solve for Expression' and 'Integrate on Variable' options can be used to generate the exact analytic form of the indefinite or definite integrals. However, they too may run into problems.

Having many seemingly alike tools to accomplish a task is an advantage. One technique may offer a solution another cannot.

We will examine the strengths and weaknesses of these analytic and numeric tools and explore alternate techniques of numerical integration.

Warmup

Symbolic and Numeric Integrals

Mathcad's subset of Maple allows you to determine a 'closed' or 'symbolic' form of the antiderivative for a wealth of functional forms. The output is equivalent to the result for an indefinite integral up to the integration constant C.

However, not all integrands have analytic closed-form solutions. The impressive array of analytic integration techniques and shortcuts simply won't allow one. As long as the integrand is continuous over the limits of integration, Mathcad's numerical techniques can take over and output an estimate of the definite integral.

For the function $f(x) = x^2$, the Symbolic Processor returns the antiderivative after you have selected the variable x from the *expression* of the integrand (not from the integral) and, under the Symbolic menu, have selected 'Integrate on Variable'. The expression returned, as shown in Figure 1, is the exact analytic form and can be cut, copied and pasted into any Mathcad expression.

$$x^2 \qquad \text{by integration, yields} \qquad \frac{1}{3}\cdot x^3$$

Figure 1

To determine a definite integral over a well defined range, a full integration expression must be created and selected. Although you may be tempted to select only the integration variable and use the same 'Integrate on Variable' option as before, the answer generated would be open to some confusion. The result is shown in Figure 2.

$$\int_3^4 x^2 \, dx \qquad \text{by integration, yields} \qquad \frac{37}{3}\cdot x$$

Figure 2

Simple analysis shows the fractional part as correct. However, the 'Integrate on Variable' option has included an extraneous factor x in its output. By selecting the whole of the definite integral expression and using the 'Evaluate Symbolically' option from the 'Symbolic' menu, an exact answer (Figure 3) is returned.

$$\int_3^4 x^2\, dx \qquad \text{yields} \qquad \frac{37}{3}$$

Figure 3

If we now use Mathcad's numerical routine to generate an output for the same definite integral, a decimal output is generated rather than an exact fractional form.

$$a) \qquad \text{intf}(x) := \int_3^x t^2\, dt \qquad\qquad \text{intf}(4) = 12.333$$

$$b) \qquad \int_3^4 x^2\, dx = 12.333$$

Figure 4

There is an essential difference in the methods used in Figures 3 and 4 although at first glance the answers seem to be expressing the same result.

In the case of the Symbolic Processor, the antiderivative of the integrand was first determined. The upper and lower limits were then substituted into the expression and the difference taken. The process follows the definition of the definite integral as laid out in the Fundamental Theorem of Calculus.

In the case of Mathcad's numerical integration routine, no attempt has been made to solve for the antiderivative. Rather, the integrand x^2 has been used directly in a summation routine over the limits of integration. An approximation has been returned based solely on a numerical process.

The next example (Figure 5) will make this distinction clearer. Here, we examine a more complicated function but one which has a closed analytic form.

$$e^{-4 \cdot x} \cdot \cos(3 \cdot x) \qquad \text{by integration, yields} \qquad \frac{-4}{25} \cdot \exp(-4 \cdot x) \cdot \cos(3 \cdot x) + \frac{3}{25} \cdot \exp(-4 \cdot x) \cdot \sin(3 \cdot x)$$

Figure 5

We now evaluate the definite integral using 'Evaluate Symbolically' from the 'Symbolic' menu. The values of the upper and lower limits are substituted directly into the antiderivative expression and the exact difference taken.

$$\int_0^4 e^{-4 \cdot x} \cdot \cos(3 \cdot x)\, dx \qquad \text{yields} \qquad \frac{-4}{25} \cdot \exp(-16) \cdot \cos(12) + \frac{3}{25} \cdot \exp(-16) \cdot \sin(12) + \frac{4}{25}$$

Figure 6

If a numeric (decimal) answer is desired, the output can be copied to an open region and selected. With the addition of the = sign to the right of the selected region and a press of F9, a decimal output is generated. Figure 7 gives the result of this operation along with Mathcad's numerical result.

a) $\dfrac{-4}{25} \cdot \exp(-16) \cdot \cos(12) + \dfrac{3}{25} \cdot \exp(-16) \cdot \sin(12) + \dfrac{4}{25} = 0.16000$ symbolic output

b) $\displaystyle\int_0^4 e^{-4 \cdot x} \cdot \cos(3 \cdot x)\, dx = 0.16$ numerical evaluation

Figure 7

Are there forms of the integrand which are not suited to analytic solutions? Of course there are.

Using the Symbolic Processor, try to find the antiderivative of the expression $\dfrac{x^3}{\cos(x) - 3}$.

The function is continuous yet an attempt to find the closed form using 'Integrate on Variable' comes up empty. If a closed form does not exist, using 'Evaluate Symbolically' for the definite integral will not work either. The only recourse is to use the numerical summation process. For the definite integral, Mathcad returns an answer provided there are no singularities over the range of limits.

$$\int_0^2 \frac{x^3}{\cos(x) - 3}\, dx = -1.333939$$

Figure 8

Numerical Processes—The Riemann Sum

One of the many numerical processes which can be used to estimate the area under the curve for a continuous function is the Riemann sum.

A summation is made over a large number of rectangular areas which subdivide the area under the curve bounded by the function, the *x*-axis and the limits. The widths of the rectangles (Δx) are defined by the distance between successive values of *x* taken from within the lower to upper limit range. The rectangular heights are defined by the value of the function at a point within the interval of each Δx.

The base of each rectangular can be of varying width and the height of each rectangle can be defined for any point within the width of the respective base. However, the use of a common width with a clearly defined evaluation point is most easily programmed. With a common base width Δx, the height can be defined as $f(x)$ taken at the left or right endpoints of the interval or at its center.

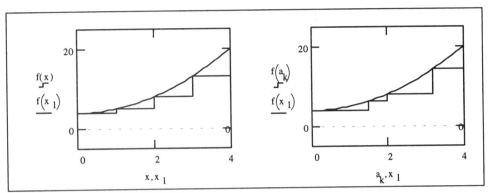

Figure 9

Figure 9 shows two rectangular breakdowns of the function $f(x) = x^2 + 4$ over the limits 0 to 4. The left-hand plot shows a regular Δx of 1 with the heights calculated at the left endpoints $f(0)$, $f(1)$, $f(2)$ and $f(3)$. The right-hand plot shows the same function with 4 rectangles of varying width. Both methods provide estimates of the area which are less than the exact area.

By defining a step size along the x-axis as the width of each base and by letting a vector index k cover the range of intervals between the integration limits, the midpoint of each of the N rectangles can be defined by $midpoint_k = \dfrac{x_k + x_{k+1}}{2}$ where the base width is given by $\Delta x = x_k - x_{k-1}$ for all $k = 1$ to N. Then the left-hand limit is defined by x_0 while x_N defines the right-hand limit.

☑ While the index varies from 1 to N, the initial value of x, (x_0), can be defined on its own. This technique solves any "index out of bounds" error messages.

The height of each rectangle for the function $f(x)$ is defined by $f(midpoint_k)$.

The sum of these areas, $\displaystyle\sum_n f(midpoint_k) \cdot \Delta x$ over N rectangles, is the Riemann sum.

Figure 10 shows a plot of the function $f(x) = x^5$ overlaid with 5 rectangles of width $\Delta x = 0.2$. The Riemann sum is given at the bottom along with the numeric answer for comparison.

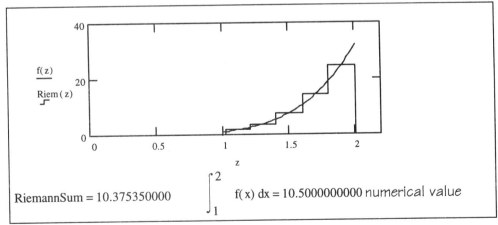

$$RiemannSum = 10.375350000 \qquad \int_1^2 f(x)\,dx = 10.5000000000 \text{ numerical value}$$

Figure 10

The use of rectangles provides a reasonable approximation to the shape of the curve. An increase in the number of panels to 20, 50 or 100 would simulate the limit process as the number of rectangles moves toward infinity while the rectangular width grows infinitesimally small. See Figure 11 for $N = 50$. The value of the rectangular area is approaching the value of 10.5 as the number of panels grows.

Other summation techniques use trapezoidal areas or quadratic function approximations between selected points on the curve.

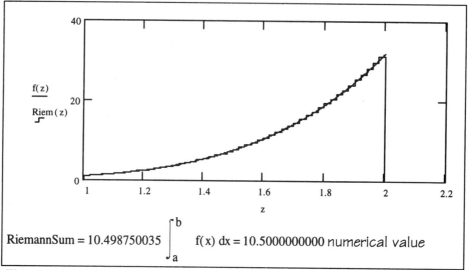

$$\text{RiemannSum} = 10.498750035 \quad \int_a^b f(x)\, dx = 10.5000000000 \text{ numerical value}$$

Figure 11

Explorations

The Basics

1. Use either the Symbolic Processor or the numerical integrator to determine the following definite and indefinite integrals:

a) $\displaystyle\int_{-3}^{3} 0.75x^4 + \frac{3x^3}{2} + 2\, dx$

b) $\displaystyle\int \cos\!\left(3\alpha + \frac{\pi}{4}\right) d\alpha$

c) $\displaystyle\int_{0}^{2\pi} \frac{1}{3}\sin^2(2x)\, dx$

d) $\displaystyle\int_{a}^{b} \frac{x^2}{x - a + 1}\, dx$

e) $\displaystyle\int_{-4}^{4} \sqrt{49 - 2x^2}\, dx$

f) $\displaystyle\int_{0}^{2.5} 20(1 - e^{-2t})\, dt$

g) $\displaystyle\int_{0}^{3} \tan^2(x)\, dx$

h) $\displaystyle\int_{4}^{5} \frac{\ln(3x)}{2}\, dx$

2. Show that the natural logarithm can be defined as $\ln(x) = \displaystyle\int_{0}^{x} \frac{1}{t}\, dt$ for $x > 0$. Compare the accuracy of this definition to that of the built-in function 'ln'.

Beyond the Basics

1. Develop the Riemann sum technique for a function of your choice. There are many possible variations of the technique although the three clearest are the use of the left, right and center of the interval for the evaluation of the rectangular height. If you can program one of these, the others are just an edit away. Compare each method for its under-estimate or over-estimate of the definite integral. Can the left and right endpoint methods be used to bound the exact value?

2. The most basic form of numerical integration is the 'average ordinate' method. Since the average of a function y over $[a, b]$ is defined as $y_{avg} = \dfrac{1}{b-a}\displaystyle\int_a^b y\,dx$ then the integral of the function over $[a, b]$ is given by $\displaystyle\int_a^b y\,dx = (b-a)y_{avg}$. Determine the approximate area under the curve $y(x) = \dfrac{e^{-x}}{\cos(x)}$ evaluated at six points between $x = 0$ and $x = 1$ (e.g., $x = 0, 0.2, 0.4, 0.6, 0.8$ and 1.0). Compare the calculated value to the result of definite integration over the same range. Does the difference between the two methods diminish as the number of points increases?

3. Develop a integration-summation routine which uses the trapezoidal approximation to the function and apply it to any continuous function. The area of a trapezoid is given by $\dfrac{1}{2}(\Delta x)[f(x\_left) + f(x\_right)]$ where $f(x\_left)$ and $f(x\_right)$ represent the respective lengths of the parallel and unequal sides.

4. Research Simpson's Rule for numerical integration and develop a Mathcad file which clearly shows the process and its output for a function of your choice. Simpson's Rule assumes the function displays a roughly parabolic form between points along the curve.

8.9 Applications of the Definite Integral

The study of dynamic systems is permeated by derivatives and integrals. But, why bother with all this formalism, all these rates and summations? Unfortunately, truly static systems (i.e., free of calculus) are virtually non-existent. Even the expression 'static system' is on a par with oxymorons such as 'eternal instant'. Something as statically overwhelming as an office tower undergoes shifts in response to forces exerted upon it and displays dynamic behavior continuously.

Definite integrals then appear in the languages of electrical circuit theory, mechanics, audio and electromagnetic wave theory, quantum physics and sound signals. The list goes on and on.

The integrals we have examined so far may seem removed from the purpose of calculus: to solve differential equations, those descriptions of the dynamic behavior of systems. However, even in the integration of a simple expression such as $\int_0^x t^2 dt$, the differential equation $y' = x^2$ is being solved with the initial condition $(0, 0)$. The techniques of integration implicitly contain the study of differential equations.

Warmup

Mechanics

In linear mechanics, the inertia or mass of a body is a measure of its resistance to the applied forces. Newton's Second Law of Motion, $F = ma$, defines the mass as the ratio of the applied force to the resulting acceleration.

The moment of inertia is a measure of the resistance of a mass to rotational motion about an axis and is defined as $I = \sum_i m_i r_i^2$ where i represents an index over all small masses m_i located a distance r_i from the rotation axis.

A circular disk of radius R can be thought of as being made of many concentric rings. If the disk rotates about a perpendicular axis passing through the center, the part of the total moment which any ring contributes is given by $\Delta I = r^2 \Delta M$. Here, r is the radial distance from the center of the disk to the inner diameter of the ring while ΔM is the mass of the ring. For M as the total mass, the increment contributed by the ring is given by $\Delta M = \dfrac{2r\Delta r}{R^2} M$.

In general, for a disk of mass M and radius R, the moment of inertia is $I = \int_0^R \dfrac{2M}{R^3} r^3 dr$.

An evaluation of this can be made for the general case (the indefinite integral) or the specific case (the definite integral). Once this example file has been created, it can be edited for other disk sizes or for any shape.

Double Integrals and the Use of Symmetry

In the case of the moment of inertia for a disk, an integration over the 2-dimensional surface has been replaced with an integration process over one variable. We have used the symmetry of the disk to reduce the complexity of the integration.

Both Mathcad and the Symbolic Processor can integrate over more than one variable by embedding one integral within another. Figure 1 shows the result for the symbolic integration of the function $f(x, y) = x^2 + y^2$ over a region of the x-y plane.

$$\int_c^d \left(\int_a^b x^2 + y^2 \, dx \right) dy \qquad \text{...integrates symbolically to...}$$

$$\frac{1}{3} \cdot b^3 \cdot d + \frac{1}{3} \cdot b \cdot d^3 - \frac{1}{3} \cdot a \cdot d^3 - \frac{1}{3} \cdot a^3 \cdot d - \frac{1}{3} \cdot b^3 \cdot c - \frac{1}{3} \cdot b \cdot c^3 + \frac{1}{3} \cdot a \cdot c^3 + \frac{1}{3} \cdot a^3 \cdot c$$

Figure 1

For the inner integral, the variable y is considered a constant and only integrated when the limits a and b have been substituted into the result of the inner integration. The numeric process can confirm that this extended expression is a proper representation of the integral. Figure 2 shows the result of a nested numeric integration and of the confirmation upon substitution of the values a, b, c and d into the symbolic expression.

$$\int_0^2 \left(\int_0^3 x^2 + y^2 \, dx \right) dy = 26 \qquad \text{numerical evaluation}$$

$$a := 0 \quad b := 3 \quad c := 0 \quad d := 2 \quad \text{confirmation of indefinite integral}$$

$$\frac{1}{3} \cdot b^3 \cdot d + \frac{1}{3} \cdot b \cdot d^3 - \frac{1}{3} \cdot a \cdot d^3 - \frac{1}{3} \cdot a^3 \cdot d - \frac{1}{3} \cdot b^3 \cdot c - \frac{1}{3} \cdot b \cdot c^3 + \frac{1}{3} \cdot a \cdot c^3 + \frac{1}{3} \cdot a^3 \cdot c = 26$$

Figure 2

Instead of integrating over small surface areas of size *dxdy*, the radial symmetry in the disk reduces the process to integration over small changes in the radius *dr*.

The same reasoning can be applied to volumes which display cylindrical or spherical symmetry. By a proper choice of the integration variable, the 3-dimensional problem can be reduced to a two- or one-dimensional one.

In the case of the volume of a sphere, we could sum over all small cubic volumes *dxdydz* within the sphere. However, the volume can be sliced into disks of area πr^2 and thickness Δz. As an egg can be rebuilt after passing through an egg-slicer, the identity of the sphere remains intact since the area takes into account two of the three dimensions and z takes into account the third.

Since the radius of the disc (as measured in the x-z plane) is x for a given height z above the x-y plane, the whole integral can be described by $VOL = \int_{-R}^{R} \pi (r^2 - z^2) dz$. For the sphere shown in Figure 3, the disk slices can be seen as a series of latitude rings of infinitesimal thickness.

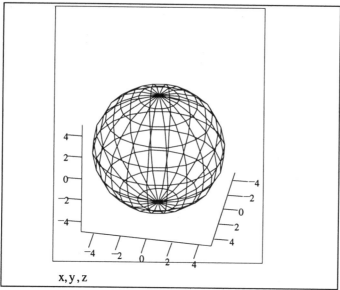

x, y, z

Figure 3

For a sphere of radius 5 units, the total volume generated by the sliced disks is given in Figure 4(a). Figure 4(b) shows that the generalized process yields the accepted equation for the volume of a sphere.

$$a) \int_{-5}^{5} \pi \cdot (25 - z^2) \, dz = 523.599$$

$$b) \int_{-R}^{R} \pi \cdot (R^2 - z^2) \, dz \quad \text{yields} \quad \frac{4}{3} \cdot \pi \cdot R^3$$

Figure 4

The process of integration is not limited to three dimensions. The volumes of objects are one particular application in our 3-dimensional world of (length units)^3. A physical system may need many more variables to describe its state and integrations can occur over these numerous dimensions.

Explorations

The Basics

1. A heat exchanger is constructed of an elliptical tube surrounded by a cylindrical tube of radius 5.0 cm. Determine the cross-sectional area between the tubes if the outer is defined by $x^2 + y^2 = 9.5$ while the inner tube is defined by $\dfrac{x^2}{1.00} + \dfrac{y^2}{2.25} = 1$. This problem can also be solved by simple analysis.

2. In the study of Mechanics, the amount of work done on a system is the sum of the product of the applied force and the distance the force was applied through in the direction of the force. For a force applied to an object, the sum tends toward the integral, $W = \int F dx$, if the mass moves in the direction of the force. The restoring force exerted by a spring is

given by Hooke's law, $F = -kx$, where k is the spring constant. Determine the amount of energy (work done on a system) needed if the spring is pulled x cm from equilibrium. Show both a general and a specific solution. The units of force are Newtons (N), the units of distance are meters (m) and k, the spring constant, is in N/m.

3. The average of a continuous function $y(x)$ over [a, b] is defined as $y_{avg} = \dfrac{1}{b-a}\displaystyle\int_a^b y(x)dx$

   while the RMS (root-mean-square) value for the same function (or signal) is

   $y_{RMS} = \sqrt{\dfrac{1}{b-a}\displaystyle\int_a^b [y(x)]^2 dx}$. Determine the average and RMS values for the following

   sinusoidal signal: $i(t) = 10.0 \sin(377t - 30°)$ mA over a time interval of your choice.

4. An aluminum pipe fitting in designed by rotating the first quadrant area of the function by 360° about the z axis between the lines $z = 2$ cm and $x = 1$ cm. Determine the mass of the fitting if its density is 2.70 g/cm³.

Beyond the Basics

1. The cross-section of a cam centered at the origin is created from the intersection of an ellipse and a hyperbola. Determine the area of the cam face (in cm²) if the boundaries

   are given by $\dfrac{x^2}{25} + \dfrac{y^2}{4} = 1$ and $\dfrac{(x+5)^2}{9} - \dfrac{y^2}{4} = 1$ with x and y in centimeters.

2. Use the definition of the moment of inertia of a disk given in the Warmup to determine the moment of inertia of an elliptical plate about a perpendicular axis through its origin.

3. Waveforms (sound, light, water) can be decomposed into superpositions of sine and cosine waves. These Fourier components are the pure waveforms underlying the wave. The amplitude of each harmonic (sinusoidal) component can be determined from an integration process which measures the overlap between the original waveform and its compo-

   nents. For a periodic square wave given by $square(x) = \begin{cases} -1 \text{ for } 0 \le x < \pi \\ +1 \text{ for } \pi \le x < 1 \end{cases}$ over a cycle of

   [0, 2π], the coefficient of the nth sine function harmonic is

   $b_n = \dfrac{1}{\pi}\displaystyle\int_0^{2\pi} square(x) \cdot \sin(nx)dx$. Determine the general form for the amplitude of the nth

   harmonic. Generate the specific numerical results up to $n = 7$. Show a plot of the harmonic amplitude versus the harmonic number. This is the frequency spectrum of the signal.

4. Repeat the method of #4 for the periodic sawtooth function given by $f(x) = \dfrac{x}{\pi} - 1$ over

   [0, 2π] up to the Nth harmonic (your choice). Plot the sum of the generated harmonic

   functions, $sum(x) = \displaystyle\sum_{n=1}^{N} b_n \sin(nx)$ and compare this waveform to the original function.

   What do you notice about the original and the synthesized waveforms as the number of harmonics in the sum is increased?

5. Determine the volume of an ellipsoid (general and specific). An ellipsoidal volume is created from the rotation of an ellipse in the x-z plane by 360° around the z-axis.

Chapter 9: Series Approximations

Even the simplest scientific calculator allows you to perform basic arithmetic operations and to readily determine the values of all of the trigonometric, exponential and logarithmic functions. For the arithmetic processes, the numbers input into the calculator are the elements used by an internal machine arithmetic. The generated answers are output to the display.

How then can this same calculator determine the sine of all angles? or the logarithm of any number (excluding zero and negative numbers)? By definition, the trigonometric functions are the ratios of two sides of a right-angled triangle. And the logarithm of a number is the power to which the base must be raised to represent that number. The calculator does not have enough memory to store the necessary information for every possible variation of the angle and the lengths of the sides of the triangle you may be interested in or for every possible number expressed as a base raised to a power.

And yet, the speed at which a calculator returns a result would have been thought a miracle by those who had to calculate sines, cosines, tangents and logarithms using involved routines or by those who survived having to look up "accurate" values of the trigonometric and logarithmic functions in extensive tables. The problem at hand became secondary to the time taken to find appropriate values.

The solution to a calculator's lack of infinite memory: approximation. The ratio of two sides of a triangle can be approximated as an finite summation series whose elements are powers of the input angle. No ultimate, infinite memory bank is needed to store all possible trigonometric outputs of all possible angles. Logarithms and powers can be estimated using the summation of simple terms. As long as the approximation is accurate enough for your needs, this method works. And, as in most scientific and mathematical situations, if the system works, it is useful.

The first set of exercises in this chapter will introduce you to the Taylor Series, a polynomial series approximation used in estimating values of trigonometric and transcendental functions given that there exists knowledge of the function and all its derivatives at one point within the domain of interest.

The remainder of the exercises will focus on the Fourier Series, a series approximation which uses trigonometric functions to simulate regular, repeating waveforms. The concept of the Fourier Series is a powerful tool in the synthesis, analysis and filtering of waveforms.

Both of these elegant solutions have their drawbacks. The Taylor Series can quickly diverge from the intended result dependent on the number of terms included in the expansion and the position of the reference point relative to the point of interest. As well, the assumption is that the series contains an infinite number of terms. In practical terms, this is impossible. With the Fourier Series, the assumption of a sine wave of infinite extent creates a problem given that all synthesized and analyzed signals are finite in extent. As well, the inclusion of only a finite number of terms within the expansion leads to a ringing effect at points of transition.

9.1 Taylor Series

The question of series approximations is not a new one.

Zeno's (Greek mathematician, ca. 450 B.C.) paradox of Achilles and the Tortoise was one of the earliest recorded problems addressing the issue of the bounded sum of an infinite number of successively smaller terms. Achilles gives the Tortoise a head start and, even though Achilles is a swifter runner than the Tortoise, he must always catch up to where the Tortoise has just been. The Tortoise is then always ahead.

We know this situation is absurd from our experience.

The school of mathematician/priests led by Pythagoras was the reigning mathematical school of the day. The Pythagoreans held to a doctrine that "Numbers constitute the entire heaven" or "All is Number". They believed in the existence of the geometric point and held multiplicity and change to be universal truths.

Zeno, on the other hand, was a member of the Eleatic school of philosopher/mathematicians. The school's founder, Parmenides of Elea, supported ideas which were in direct opposition to those held by the Pythagoreans. Zeno's paradox of Achilles and the Tortoise uses the Pythagorean idea of divisibility applied to distance to show the absolute absurdity of such a notion. If distance was divisible then motion was impossible. And since objects do move then the Pythagoreans must be in error.

Zeno created other paradoxes of time and space and, working from his opponents' own premises, would show the error of the initial statements.

The absurdity of the Achilles and the Tortoise paradox may seem obvious and trivial to you. You may even wonder how anyone could possibly have wasted any time thinking such nonsense. However, we exist in a calculation age. The manipulation of numbers is familiar to most people in our society and is an integral part of our culture.

The Greek philosophers of Zeno's era were language oriented. They spoke rather than calculated. And the problem of division, of infinitesimals and remainders, was far more difficult for them than any other arithmetic operation. In fact, they had no way of doing division to *any order of accuracy*.

The numerical culture of Zeno's time accepted the notion of adding successively smaller amounts to a given amount. The expected result was the continued and eventual unbounded growth of the quantity. As an example, the infinite addition of progressively smaller grains of sand to a pile would result in an infinitely large pile given enough time (and enough sand).

In this set of explorations, we will see that this 'expected' result is not necessarily the case.

The use of series approximations brings up the related questions of exactness, accuracy and precision. If you are using a calculator to determine the sine of an angle and that calculator is using an approximation routine to offer you a display of the result, does an 'accurate' value exist? And, if an infinitely precise or exact value does exist, what does it look like? The calculator cannot store every possible instance of the sine function nor can it store all possible ratio combinations of a right-angle triangle.

This is not a trivial problem. Calculations of real, physical data are initially limited by their inherent reading errors, whether human or machine. Added to this problem is any computational error that may be introduced along the way to the output of a result.

Welcome to the fine balancing act of *accuracy* versus *time*. Infinite numerical accuracy would require an infinite amount of computational time. It would paralyze us with its demands. Thus, we make do with less-than-perfect knowledge so that we may solve, probe, create... and keep breathing.

Warmup

The Taylor Series

The creation of the Taylor Series for a function involves an analysis and knowledge of the function's amplitude and the value of its derivatives at *one value* of the variable *x*. Within a restricted range (the interval of convergence), the behavior of the function at other points can be projected from this 'seed' point. The error associated with using only a finite number of terms within the series can usually be made less than the desired tolerance.

☑ On an historical aside, the series developed by Brook Taylor (England, 1685-1731) was said to have been plagiarized from a similar series of Jean Bernoulli (1667-1748). However, both mathematicians' work was predated by a series developed by James Gregory (Scotland, 1638-1675).

The series in *x* expanded about the point *a* can be expressed as:

$$f(x) = \sum_{n=0}^{\infty} \frac{f^{(n)}(a)}{n!}(x-a)^n$$

where $f^{(n)}(a)$ represents the *n*th derivative of *f*(*x*) evaluated at the point *x* = *a*. This series assumes the index *n* grows to infinity. In practical terms, an infinite series is not calculable. There is enough sand. There is just not enough time or patience. Thus the function *f*(*x*) expanded at *x* = *a* can be expressed as:

$$f(x) = P_n(x) + R_n(x)$$

where $P_n(x)$ is the *n*th degree Taylor polynomial or *n*th partial sum of *f*(*x*) at *a* and $R_n(x)$ is the associated error. The finite series then is broken down into a part useful in further calculations and a remainder which agrees within the needed accuracy of the result.

$$P_n(x) = f(a) + \frac{f'(a)(x-a)}{1!} + \frac{f''(a)(x-a)^2}{2!} + ... + \frac{f^{(n)}(x-a)^n}{n!}$$
$$R_n(x) = \frac{f^{(n+1)}(\xi)(x-a)^{n+1}}{(n+1)!}$$

Within the expression for $R_n(x)$, ξ is a number on the open interval (*a*, *x*).

The expression of a function in terms of a series of derivative values may seem obscure. If the derivative gives the necessary local information about the function's rate of change, then this approach makes sense as the amplitude *f*(*a*) gives an indication of where the function is at *x* = *a* while the derivatives provide information on where the function is going as *x* leaves *a*. The first derivative provides the immediate slope; the second derivative provides an idea of the curvature of the function or the rate of change of the rate of change and so on.

The amplitude $f(a)$ indicates where the function is; the derivative, $f'(a)$, indicates where the function is going.

To obtain an appreciation of the elegance of the Taylor series approach, let's examine the sine function around the origin, for $a = 0$. For $n = 0$ to N, the function $f(x) = \sin(x)$ can be approximated around the origin by

$$f(x) = x - \frac{x^3}{3!} + \frac{x^5}{5!} - \frac{x^7}{7!} + \ldots$$

Within the expansion, the even powers of x have coefficients which reduce to zero (i.e., $\sin(0)/n!$) while the odd powers have coefficients of $\cos(0)/n!$ with alternating signs. Using the summation symbol and a generalized expression for the series term,

$$f(x) = \sum_{k=1}^{0.5(N+1)} \frac{(-1)^{k+1} x^{2k-1}}{(2k-1)!}$$

Notice while the index n covers from 0 to N, the generalized expression in k (from $k = 1$ to $k = \frac{1}{2}(N+1)$) includes only odd powers of x.

The remainder for the Nth degree Taylor polynomial for any value of x would be given by

$$R_N = \frac{f^{(N+1)}(\xi) x^{N+1}}{(N+1)!} \quad \text{where } \xi \text{ is between 0 and } x.$$

We can compare our estimates to the 'exact' value of $f(x)$ defined by Mathcad's sine function accurate to 15 decimal places. We may also visually examine the behavior of the Taylor polynomial as its degree is increased.

In Figure 1, a general term is defined for a particular angle. The exact numerical value of the sine accurate to 15 decimal places is included. Figure 2 compares successive summations of the general term to the numerical value of the sine function at that point.

Taylor Series for the SINE function expanded about a = 0

A generalized expression is used and successive partial sums are compared to the accurate value of the sine function at 'x'.

$x := \dfrac{\pi}{7}$ ... value of angle in radians

$N := 10$ ... the number of even-power expansion terms for the partial sum

$k := 1 .. N$ ... the index of the expansion term

$u_k := (-1)^{k+1} \cdot \dfrac{x^{2 \cdot k - 1}}{(2 \cdot k - 1)!}$ ... general k^{th} term, expansion of sine function

$\sin(x) = 0.433883739117558$ ... exact value of sine function at 'x'

Figure 1

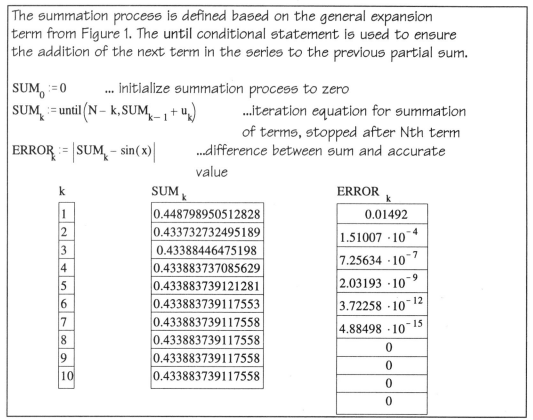

The summation process is defined based on the general expansion term from Figure 1. The *until* conditional statement is used to ensure the addition of the next term in the series to the previous partial sum.

$\text{SUM}_0 := 0$ ... initialize summation process to zero

$\text{SUM}_k := \text{until}\left(N - k, \text{SUM}_{k-1} + u_k\right)$ ...iteration equation for summation of terms, stopped after Nth term

$\text{ERROR}_k := \left| \text{SUM}_k - \sin(x) \right|$ ...difference between sum and accurate value

k	SUM_k	ERROR_k
1	0.448798950512828	0.01492
2	0.433732732495189	$1.51007 \cdot 10^{-4}$
3	0.43388446475198	$7.25634 \cdot 10^{-7}$
4	0.433883737085629	$2.03193 \cdot 10^{-9}$
5	0.433883739121281	$3.72258 \cdot 10^{-12}$
6	0.433883739117553	$4.88498 \cdot 10^{-15}$
7	0.433883739117558	0
8	0.433883739117558	0
9	0.433883739117558	0
10	0.433883739117558	0

Figure 2

For example, an examination of the tables reveals that for $k = 7$ (a power of x^{13}) the approximation matches the accurate value to within 1 part in 10^{15}.

A plot of this partial sum versus the sine value for the first 10 non-zero terms of the expansion (or up to x^{19}) is given in Figure 3. The series expansion converges quickly for the chosen value of x. Slight fluctuations above and below the accurate numerical value can be examined by using the Zoom function in the X-Y Plot menu or by restricting the plot range more tightly about the exact value. Figure 4 shows an expanded view of the sum plot for $k = 4$ to 7.

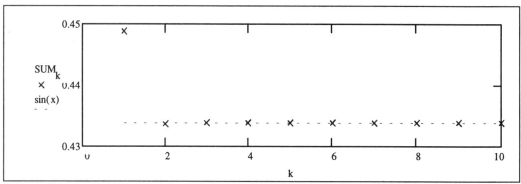

Figure 3

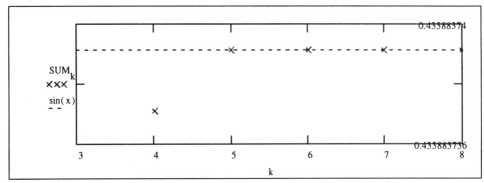

Figure 4

In the next example, the **until** function is used to halt the series after the requisite tolerance (a match to within the error) is achieved. The initial values are the same as those given in Figure 1.

$$\mathrm{SUM}_k := \mathrm{until}\left(\left|\sin(x) - \mathrm{SUM}_{k-1}\right| - \mathrm{error}, \mathrm{SUM}_{k-1} + u_k\right)$$ iteration eq'n for summation of terms with error= 10^{-10}

$\mathrm{N2} := \mathrm{last}(\mathrm{SUM}) - 1$ convergence test
$\mathrm{N2} = 5$ limited by 'until' function
$j := 1 .. \mathrm{N2}$ and specified error---halted
     after N2 terms

tables of index and terms to N2

j	SUM_j	$\left\|\sin(x) - \mathrm{SUM}_j\right\| - \mathrm{error}$
1	0.448798950512828	0.014915211295269
2	0.433732732495189	0.000151006522369
3	0.43388446475198	0.000000725534421
4	0.433883737085629	0.000000001931929
5	0.433883739121281	- 0.000000000096277

Figure 5

The previous calculations have all taken place at a particular value of the angle. The variable x is now allowed to vary over its entire domain rather than be defined at a single point. If we explore the Taylor series approximation to the sine function using comparative plots, we see that there is a reasonable match over a restricted range. This range defines the usefulness of the expansion to the predefined number of terms.

For $n = 0$ to 5, the sine function would be approximated by $f(x) = x - \dfrac{x^3}{3!} + \dfrac{x^5}{5!}$

In Figure 6, the range of x values for which there is a reasonable match is approximately given by $(-2, +2)$.

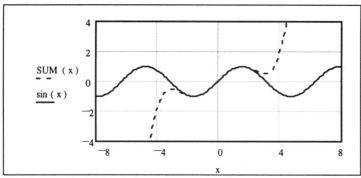

Figure 6

If the number of terms in the expansion is increased, the area of overlap is extended. Figure 7 represents the series approximation to the sine function for $k = 0$ to 17. The sine function seems to behave nicely, the area of overlap between the approximation and the function increasing as the number of terms in the expansion is increased.

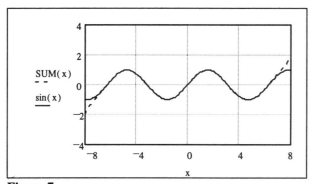

Figure 7

The overlap of the approximation now extends almost to the extremes of the plot window along the *x*-axis. Even though the overlap at $x = 4$ appears to be perfect, an examination of the calculated numbers in that area still shows there to be a small error. Figure 8 shows the ERROR(x) function for the expansion of sine with $n = 0$ to 17.

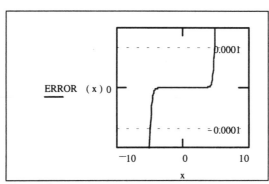

Figure 8

Explorations

Helpful Mathcad Utilities—A Reminder

a) Series approximations

The series approximations to any of the above functions can be found using the Symbolic Processor. Define the function as you normally would. Select the variable. Then select Symbolic/Expand to Series from the menu and specify the highest order of the expansion.

b) Summation (save your fingers)—see also Section 3.5, page 61.

Each of the terms in the series can be expressed using a generalized index n and a generalized expression, u_n. In this way, the partial sum up to the nth term can be expressed using the summation symbol Σ. This facilitates the calculation by allowing an iterative process to define the partial sum rather than having to define the partial sum for every possible value of n and repeatedly change the highest index number. The Σ symbol is invoked by clicking on the summation symbol within the vertical palette in version 5.0, by selecting from the Calculus Palette in version 6.0 or by typing $.

c) Conditional statements (for limiting accuracy)—see also Section 8.3, page 143.

Mathcad is not strictly a programming language and so does not contain the conditional loops and branching statements used for iteration. These can be programmed with the PLUS versions. In the Standard versions, two utilities exist which mimic these processes: the **if** function and the **until** function.

if(conditional statement, truevalue, falsevalue)

until(condition, indexed equation)

d) Last but not least

The **last** utility returns the index of the last element in a vector and has the syntax **last**(vector_name). It is used in conjunction with the conditional statements applied to a vector process.

e) The **Zoom** utility

When a graphical region has been selected, a new menu item, **X-Y Plots**, appears along the top border. Within this menu is a **Zoom** item. When the cursor is placed within the selected plot region and dragged, a box is formed around the region to be expanded. Choosing **Unzoom** returns the original plot format.

Table of functions

$f(x)$	expanded at $a =$
$\sin(x)$	0
$\cos(x)$	0
e^x	0
$\ln(x)$	1

The Basics

1. With reference to the functions in the above table, examine the numerical behavior of the expansions at a point x as the number of terms n in the partial sum increases. Mathcad allows a precision of up to 15 decimal places. How do you define and determine the index at which the summation should end?
2. Examine the effect of the distance of the computed point x from the seed point a for the same number of terms. As a starting example, if $a = 0$, what is the difference in the series nth partial sum evaluated at $x = 1$ versus $x = 5.0$ in terms of the 'exact' value?
3. Examine the symmetry of the expanded function relative to the symmetry of the expansion terms. Would you expect the symmetry to be conserved? Can you expand odd functions by using even terms?
4. For values of x for which the series would diverge or converge slowly, how do you go about translating the expansion so as to regain convergence or, at least, allow more rapid convergence? Show your technique applied to each of the series expansions. For example, the evaluation of $\sin(7\pi/8)$ may not be manageable to within the desired accuracy from the seed point $a = 0$ with only $n = 0$ to 4 included. Shift the point to $a = 3\pi/2$ and compare the rate of convergence (the decrease of error) to that of the original series approximation.
5. Examine the behavior of the expansions as the form of the independent variable is changed. For example, how would you define the expansion for $f(x) = \sin(3x)$ knowing the form of the $\sin(x)$ expansion? Realize that x is as unknown as $3x$. Generate new series approximations and compare them to the originals:
   a) $\sin(3x)$ versus $\sin(x)$
   b) $4\,e^{-3x}$ versus e^x
   c) $2\cos(x^2)$ versus $\cos(x)$

Beyond the Basics

1. Given the expressions you have developed for the expansion of $\sin(x)$, $\cos(x)$ and e^x show for the representation of complex numbers:
$$|Z|\, e^{j\theta} = |Z|\, (\cos\theta + j\sin\theta) \text{ for all } \theta$$
where j is the complex rotation operator, $|Z|$ is the amplitude of the vector $Z = x + jy$ and θ is the related argument (or angle).
2. For the sine function, $f(x) = \sin(x)$, examine the behavior of the remainder $R_n(x)$ as n approaches infinity. How would you use the absolute value of the remainder to set the accuracy of the intended result?

References

1. *A History of Mathematics*, Boyer and Merzbach, 2nd ed, John Wiley and Sons, 1989, pp. 74-79
2. *Mathematics for the Millions*, L. Hogben, Norton and Co., 1938, pp.16-20

9.2 Fourier Series and Smoothing Functions

Jean-Baptiste Fourier (France, 1768-1830) was examining the transfer of heat across a boundary when he formulated his expansion of functions in terms of a series of trigonometric functions. Just as the Taylor Series can be used to approximate a function using power functions, the repeated superposition of trigonometric functions of infinite extent can be used to closely approximate a function, especially those which display a repetitive or cyclical behavior.

An infinite series of infinite extent poses some problems in its application to the synthesis and analysis of signals or waveforms which, by virtue of their being real, are necessarily finite. The real signals we use for communication have limited extent. However, as in the use of other approximations, if the application of the series to the synthesis and decomposition (analysis) of the waveform works to the desired accuracy, the series is useful.

And, if we are to use instruments to synthesize and analyze these signals, the computational power is limited by hardware, software and cost.

In this set of explorations we will examine the behavior of a periodic waveform and observe the interplay of the waveform with its various supposed harmonic components. In this way, the amplitude of the separate harmonics of fundamental sine and cosine functions will allow us to synthesize the waveform. We can then compare the synthesized wave to the original.

In the process of synthesis, the inclusion of a finite number of terms will create an error in the 'exact' waveform typified by ringing or overcompensation (the Gibbs phenomenon). This error will be especially prevalent near points of sudden change.

In a later set of explorations, the process will be reversed. A signal will be broken down into its discrete components, the coefficients (amplitudes) of the sine and cosine frequencies. In this a way, a frequency spectrum is generated from the analysis of the waveform. While synthesis takes the frequency domain information and transforms it into a time domain signal, the reverse is true for analysis.

Warmup

A Fourier polynomial of degree n is a series created from the superposition of fundamental sine and cosine waves and their harmonics up to and including degree n.

$$F_n(x) = \frac{a_0}{2} + \sum_{k=1}^{n} a_k \cos(kx) + \sum_{k=1}^{n} b_k \sin(kx)$$

If a function is not well-behaved and displays jumps and discontinuities, its Fourier estimate allows its differentiation due to the continuity and differentiability of the constituent trigonometric functions. When you examine functions which exhibit these defective behaviors, Fourier's insight shows itself to be powerful and effective.

We will apply the idea of a trigonometric series approximation to the synthesis of a sawtooth wave, a regular wave characterized by a periodic linear increase from a minimum to a maximum value followed by a sudden downward leap to the minimum value. Figure 1 shows a sawtooth wave of unit amplitude with a period of 2π. The lack of a perfectly vertical jump is an artifact of Mathcad. Its effect can be reduced by choosing a smaller step size.

This waveform provides a linear ramp and is used in the scanning voltages of cathode ray tubes where the trace must repeatedly move across the screen from left to right.

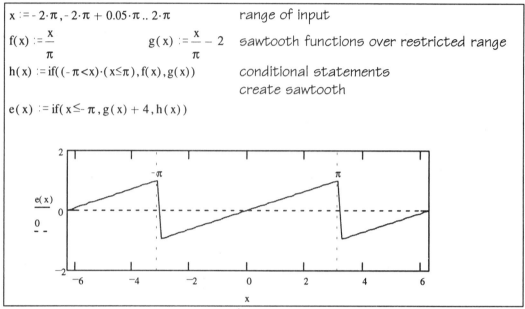

$x := -2 \cdot \pi, -2 \cdot \pi + 0.05 \cdot \pi .. 2 \cdot \pi$ range of input

$f(x) := \dfrac{x}{\pi}$ $g(x) := \dfrac{x}{\pi} - 2$ sawtooth functions over restricted range

$h(x) := \text{if}((-\pi < x) \cdot (x \le \pi), f(x), g(x))$ conditional statements create sawtooth

$e(x) := \text{if}(x \le -\pi, g(x) + 4, h(x))$

Figure 1

The theory of the discrete Fourier Series generates the recipes for the determination of the coefficients, a_0, a_k and b_k given the exact functional form of the desired synthesized wave.

For a wave of period 2π given by $f(x)$ over the domain $[-\pi, +\pi]$ then the initial coefficient (or constant term) is given by $a_0 = \dfrac{1}{\pi}\int_{-\pi}^{\pi} f(x)dx$. This value sets the baseline for the periodic behavior of the function (in Electrical theory, it can be interpreted as the DC shift).

The other coefficients of the sine and cosine frequencies are given by $b_k = \dfrac{1}{\pi}\int_{-\pi}^{\pi} f(x)\sin(kx)dx$ and $a_k = \dfrac{1}{\pi}\int_{-\pi}^{\pi} f(x)\cos(kx)dx$ respectively.

These integrals can be evaluated by analysis or numerically. In the case of well-behaved waveforms, the analysis is relatively painless and produces generalized expressions for the coefficients.

A visual examination of the integrands can offer clues as to the reasons some of the coefficients are stronger than others while others vanish in the integration process.

In the case of $k = 1$, the coefficients represent the fundamental harmonic content of the wave. Figure 2 displays plots of the integrands for the first of both the sine and cosine coefficients.

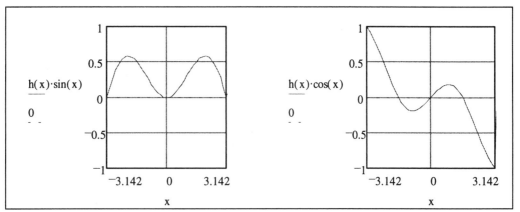

Figure 2

While the sine coefficient displays a positive integral over $[-\pi, +\pi]$, the symmetry of the cosine coefficient's integral suggests cancellation.

Figure 3 includes a plot of the second sine and cosine coefficients, b_2 and a_2 respectively, for the same increasing sawtooth waveform.

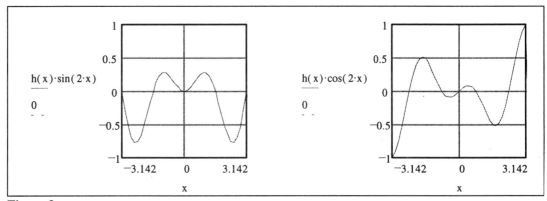

Figure 3

While the integrand for the sine function coefficient displays a larger negative contribution to the integrand this time, the cosine function coefficient's contributions once again cancel over $[-\pi, +\pi]$. The sine wave contribution seems to persist while the cosine wave contribution is nonexistent.

Would the same visual patterns exist for other values of a_k and b_k?

Examine Figure 1 and determine the value of a_0 for the sawtooth by symmetry.

A numerical analysis of the integrals is easily performed by defining generalized expressions for the integrals and by allowing Mathcad's numerical integration routine to churn through the calculations. Here, you may experiment with the Tolerance settings and the displayed accuracy of the result. However, depending on your hardware, you may have to be patient.

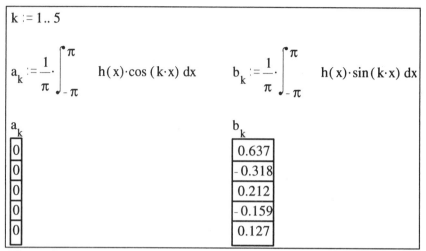

Figure 4

The numerical analysis in Figure 4 confirms the behavior of the plots in the first five harmonic components. The cosine coefficients vanish while the sine coefficients alternate from positive to negative with decreasing absolute value.

An alternative to this numerical calculation is the symbolic generation of the coefficients using Mathcad's Symbolic Processor. Figure 5 includes the generalized expression of a_k and b_k up to the fifth harmonic. By defining a variable and a range for k, tables of values can be created.

symbolic evaluation of a_0, a_k, and b_k

$$\frac{1}{2 \cdot \pi} \cdot \int_{-\pi}^{\pi} \frac{x}{\pi} \, dx \qquad \text{yields} \quad 0$$

$$\frac{1}{\pi} \cdot \int_{-\pi}^{\pi} \frac{x}{\pi} \cdot \cos(k \cdot x) \, dx \qquad \text{yields} \quad 0$$

$$\frac{1}{\pi} \cdot \int_{-\pi}^{\pi} \frac{x}{\pi} \cdot \sin(k \cdot x) \, dx \qquad \text{yields} \quad \frac{-2}{\pi^2} \cdot \frac{(-\sin(kx) + k\pi \cos(k\pi))}{k^2}$$

evaluation of sine coefficients for k = 1 to 5

$$k := 1 .. 5 \qquad b_k := \frac{-2}{\pi^2} \cdot \frac{(-\sin(kx) + k\pi \cos(k\pi))}{k^2}$$

b_k

0.637
- 0.318
0.212
- 0.159
0.127

Figure 5

These coefficients can then be used to synthesize the wave up to the fifth harmonic. Figure 5 displays the synthesized wave superimposed on its ideal function. The values used for the coefficients were those generated in Figures 4 and 5.

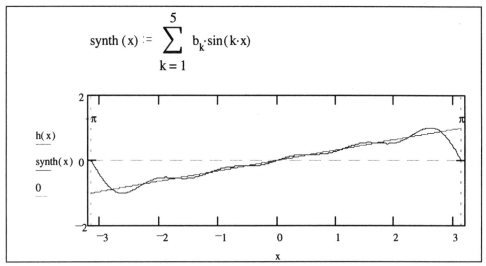

$$\text{synth}(x) := \sum_{k=1}^{5} b_k \cdot \sin(k \cdot x)$$

Figure 6

As suggested before, the coefficients fall into regular patterns for well-behaved periodic functions. And if the behavior of the synthesized wave is to be examined, the process of summing up generalized expressions is far quicker than that of calculation individual integrals. For the increasing sawtooth, the sine coefficients are given by $b_k = \dfrac{2}{\pi} \cdot \dfrac{(-1)^{k+1}}{k}$.

In the next figure the synthesis is extended to the 15th harmonic. With a generalized expression and a fine step size, there is no limit to the harmonic extent you can examine. However, saturation sets in after awhile. You get the point and the sawtooth is synthesized to the desired accuracy.

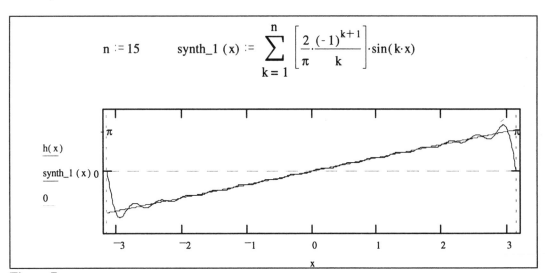

$$n := 15 \qquad \text{synth\_1}(x) := \sum_{k=1}^{n} \left[\frac{2}{\pi} \cdot \frac{(-1)^{k+1}}{k} \right] \cdot \sin(k \cdot x)$$

Figure 7

The overshooting and ringing near the jump points becomes more obvious as the harmonic content is increased. Figure 8 shows the same series for $n = 30$.

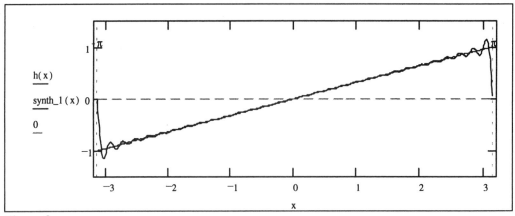

Figure 8

Although there is now a better match of the synthesized and ideal waves over the linear portion of the curve, the oscillations near the jump points are increasing in frequency and amplitude. While the inclusion of many more harmonics would reduce this effect, the process may be too time- and hardware-intensive.

One solution to the ringing problem is to contain the amplitude of each harmonic in an envelope whose purpose is to smooth out the rapid ringing. Figure 9 displays the same sawtooth with $n = 15$ overlaid by the smoothed out wave.

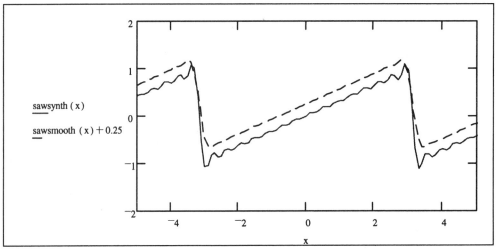

Figure 9

The smoothed function has been displaced +0.25 units vertically for clarity.

A plot of the absolute error between the exact waveform and its raw synthesized waveform is given in Figure 10. Figure 11 contains the error plot between the exact and smoothed waveforms.

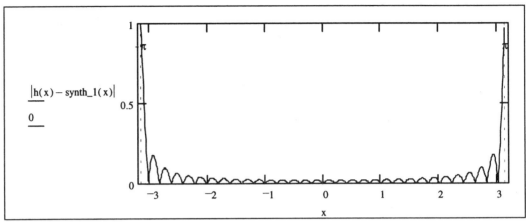

Figure 10

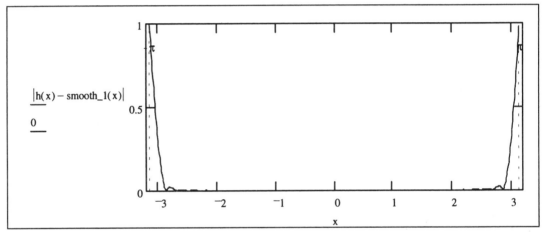

Figure 11

The variations over the linear portion have been smoothed out while an error still persists near the jump points.

The recipe for this particular smoothing effect, the Lanczos Smoothing Factor, is defined by

$$LSF(k) = \frac{\sin(\pi \cdot \frac{k}{n})}{\pi \cdot \frac{k}{n}}$$ where k is the harmonic index and n is the index of the highest harmonic

used in the series. This factor is included within the summation and alters the amplitude of each sine or cosine wave by a preset factor.

Explorations

The Basics

1. Synthesize the square wave of period 2π and unit amplitude given in Figure 12 for various numbers of harmonics. The generation of the coefficients can be treated numerically and/or symbolically.
2. Repeat the synthesis process for the triangular wave of period 2π and unit amplitude as shown in Figure 13.

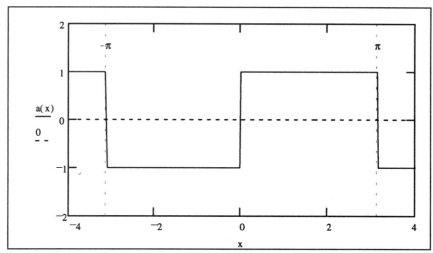

Figure 12

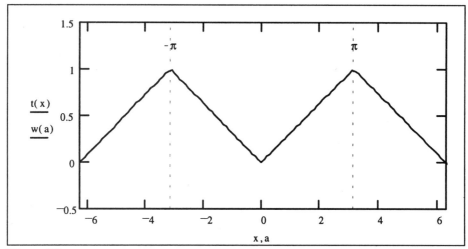

Figure 13

Beyond the Basics

1. In *The Basics* you synthesized various regular waveforms. Show, as the number of harmonics used in the series increases, that the error between the ideal function and the synthesized waveform decreases.
2. Examine the effect of applying the Smoothing Factor to either of the square, triangular or sawtooth waveforms. Compare the error with respect to the ideal waveform for smoothed and unsmoothed approximations.
3. What guidelines control the number of the final harmonic in a potentially infinite series?
4. Figure 14 contains a full-wave rectifier waveform. Synthesize this waveform and comment on the existence of only cosine functions of even frequency.
5. Are there other smoothing routines? One of these is called a Hanning function. Research its design and compare its effectiveness to that of the Lanczos Smoothing Factor.

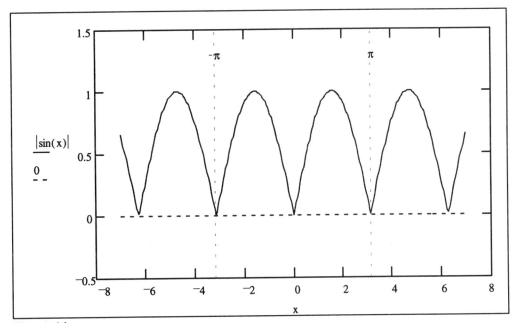

Figure 14

9.3 Fourier Sampling and the Analysis of Coefficients

The process of the Fourier Series applies in both directions, construction and de-construction. So far, we have examined the synthesis of various waveforms. Periodic functions were created according to the prescriptions offered by the coefficients of the sinusoidal harmonics within the expansions. Smoothing routines were then applied to even out the variations caused by including only a finite number of harmonics within the synthesis.

The Fourier Series can also be a tool of analysis or decomposition. Real waveforms (e.g.: water waves, radio signals) do not usually come equipped with their mathematical functions at hand. By measuring their amplitude at points along the wave, estimates of the size of their Fourier coefficients can be derived. This process will be extended in the next section which explores a method of translating the variation of the wave into its discrete frequency components.

Warmup

The first step in the process is the determination of the wave's amplitude at points on the interval. An unknown wave will be examined and its amplitude measured at discrete intervals. By associating the definite integral of a function with its average value, estimates of the coefficients of the Fourier components can be generated. These components can be checked by using them to re-synthesize the wave as a comparison to the original.

Figure 1 shows the wave and its discrete data points measured from 0 to 2π radians. Although the wave is continuous over $(0, 2\pi)$, the set of data points is discrete. For N sampling intervals over $[0, 2\pi]$, a set of $(N + 1)$ data points is created. This set of (in the case of Figure 1) 19 points represents a sample of the amplitude of the wave taken every $(2\pi/18)$ radians (or every 20°) along the horizontal axis. The finite collection of points is a representation of the continuous wave. Information between the data points is lost unless the sampling frequency is increased. However, an infinite sampling frequency would require infinite resources.

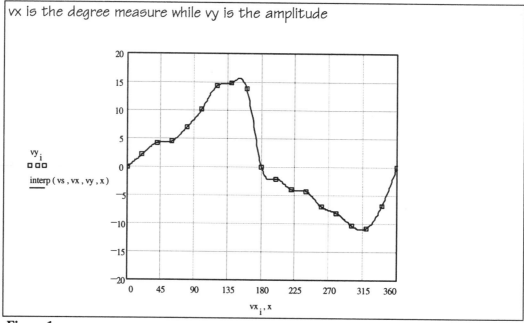

Figure 1

From the measurements of the amplitude, estimates of the coefficients are generated using the definition of the average of a function over a range in x from $[a, b]$ as $y_{AVG} = \dfrac{1}{b-a} \displaystyle\int_a^b y\,dx$.

For a cyclical function $f(x)$ over $[0, 2\pi]$, the coefficients of the cosine harmonics are given by $a_n = \dfrac{1}{\pi} \displaystyle\int_0^{2\pi} f(x)\cos(nx)dx$. If the definition of the average of a function is applied to this expression, then

$$a_n = \frac{2}{2\pi} \int_0^{2\pi} f(x)\cos(nx)dx$$

$$= 2 \cdot \frac{1}{2\pi} \int_0^{2\pi} f(x)\cos(nx)dx$$

$$= 2 \cdot [f(x)\cos(nx)]_{AVG}$$

Since the amplitudes of the wave have been measured at the sampling points, an average value of the product of the amplitude $f(x)$ and the harmonic component $\cos(nx)$ can be estimated. The same method applies for the coefficients of the sine harmonics, b_n, and the DC shift a_0.

For $N+1$ samples of (x_k, f_k) over N intervals, the values of the constant term and the coefficients are given by:

$$a_0 = \frac{2 \cdot \displaystyle\sum_{k=0}^{N} f_k}{N+1}$$

$$a_n = 2 \cdot \frac{\displaystyle\sum_{k=0}^{N} f_k \cdot \cos(nx_k)}{N+1}$$

$$b_n = 2 \cdot \frac{\displaystyle\sum_{k=0}^{N} f_k \cdot \sin(nx_k)}{N+1}$$

Figure 2 shows the data points and the analysis of the constant term and the first three coefficients of the sine and cosine harmonics for the sampled wave from Figure 1.

☑ Actually, the plot in Figure 1 was not sampled. The plot was generated from the vector of the data points input to represent a wave. However, imagine that the data points were in fact the measured results of the sampling process. The process of analysis then runs smoothly with no further tricks.

<-- Rx_k is the radian measure and f_k, the amplitude

generalized expressions for coefficients using average of the ordinates over N+1 samples

$R \cdot x_k$	$f_k :=$
0	0
0.349	2.1
0.698	4.2
1.047	4.5
1.396	7.0
1.745	10.1
2.094	14.3
2.443	14.8
2.793	13.9
3.142	0
3.491	-2.1
3.84	-4.0
4.189	-4.3
4.538	-7.0
4.887	-8.1
5.236	-10.3
5.585	-10.8
5.934	-6.9
6.283	0

$$a_0 := \frac{2 \cdot \sum_k f_k}{N+1}$$

$$a_n := 2 \cdot \frac{\sum_k f_k \cdot \cos\left(n \cdot R \cdot x_k\right)}{N+1}$$

$$b_n := 2 \cdot \frac{\sum_k f_k \cdot \sin\left(n \cdot R \cdot x_k\right)}{N+1}$$

coefficients of Fourier series: $a_0 = 1.832$

a_n	b_n
-3.953	9.57
0.222	-1.286
1.926	2.425

Figure 2

The coefficients present a coded form of the wave. They represent amplitudes of the various harmonic components and in this sense represent a frequency spectrum for the wave. The wave has then been analyzed into its component frequencies, its underlying sinusoidal content.

Here, we have only generated harmonic content up to $n = 3$. We could have very easily increased this number without end. However, at high harmonic frequencies, the period of the component waves would be far less than that of the sample interval. None of the sampled information could correspond to such high frequency variation. From our perspective as measurers of an unknown wave, there is no knowledge of the amplitude of the wave between the sampled points.

To correct this sampling shortcoming, the sampling frequency is generally chosen so as to be at twice the frequency of the highest predicted component harmonic (the Nyquist condition). In fact, over-sampling can produce an even smoother reconstruction of the wave.

Our process could then be refined by taking ever finer intervals and, necessarily, more samples. This would allow higher harmonic content in the analysis.

The test of the validity of this process is in the re-synthesis of the wave using these various calculated coefficients and the harmonics of sine and cosine. Although we expect a match if the method is consistent, we cannot expect perfect reconstruction. This would have required an infinite number of samples and this is beyond our means.

Figure 3 shows the original and reconstituted wave. Based on 19 samples and only the first few harmonics, the original has been faithfully rebuilt. Of course, if there had been any high frequency components in the original waveform, these would not have been included in the re-constituted wave.

An extreme case of this oversight can occur if the sampling frequency is far less than the general frequency components of the wave. Although the sampling data is true to *its* frequency, a reconstruction of the wave bears little resemblance to the original.

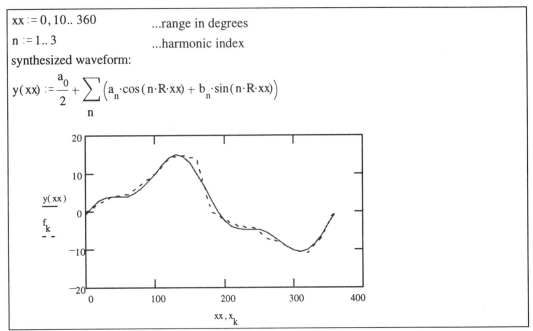

$$xx := 0, 10 .. 360 \qquad \text{...range in degrees}$$

$$n := 1 .. 3 \qquad \text{...harmonic index}$$

synthesized waveform:

$$y(xx) := \frac{a_0}{2} + \sum_n \left(a_n \cdot \cos(n \cdot R \cdot xx) + b_n \cdot \sin(n \cdot R \cdot xx) \right)$$

Figure 3

Explorations

The Basics

1. Repeat the process of analysis developed within the Warmup for the following set of data points. Set the harmonic content to $n = 3$.

amplitude	2.0	2.2	1.8	2.2	0.0	-2.2	-1.8	-2.2	-2.0
x-axis (radians)	0	π/4	π/2	3π/4	π	5π/4	3π/2	7π/4	2π

2. Compare the waveform in exercise 1 to its sampled reconstruction for harmonic content from $n = 1$ to $n = 5$.

3. Perform a sampling analysis on the following data. Set the harmonic content to $n = 5$.

amplitude	0.00	0.50	0.75	1.00	2.00	2.50	3.00	3.50	4.00
x-axis (radians)	0	π/4	π/2	3π/4	π	5π/4	3π/2	7π/4	2π

4. The points in exercise 3 represent an increasing sawtooth waveform. The waveform is well behaved and linear. For instance, other sampled points could be (π/8, 0.25) or (13π/8, 3.25). Double the number of sampled points and determine the effect of increasing the sampling frequency on the synthesis of the original waveform (keep the harmonic content in your analysis fixed).

Beyond the Basics

1. Repeat the process of analysis for the following set of data points measured over [-π,+π].

amplitude	1.00	1.13	1.28	1.45	1.65	1.87	2.12	2.39	2.72
x-axis (radians)	-π	- 3π/4	-π/2	-π/4	0	π/4	π/2	3π/4	π

a) Analyze the effect of increasing the sampling frequency at a fixed harmonic content. The equation which generated these "sampled" points is given by

$$y_k = e^{\frac{x_k + \pi}{2\pi}}$$ where k is the sampling index from 0 to N.

b) Analyze the effect of high harmonic analysis at a fixed sampling frequency.

2. Perform a sampling analysis on the smooth periodic wave over [0, 2π] defined by $y_k = \sin(x_k) + \sin(2x_k) + \cos(3x_k)$ where k is the sampling index (see Figure 2 as an example) Does the behavior of the coefficients agree with the pure sinusoidal nature of the waveform? The sampling frequency can very easily be altered by allowing k to vary over a larger range for [0, 2π].

3. Add noise to the waveform in exercise 2 by adding the term $\text{rnd}(\frac{x_k}{20})$ to the expression for y_k. This adds high frequency random noise to the otherwise smooth curve over [0, 2π]. Can the inclusion of only low frequency terms in the re-synthesis act as a filter for the superimposed noise? What are the dangers of relying on this method as a filtering process?

4. The problem of sampling frequency can be examined by comparing a waveform generated by a vector index method to the same wave generated as a continuous function. Figure 4 illustrates this problem for the function $y(x) = 2\cos(15x)$. If the sampling frequency is too low (as given by the index i applied to ff_i) then the function the sample seems to represent bears little resemblance to the original $f(xx)$. Experiment with the functions given in the figure below and determine at what point the sampling frequency provides an accurate estimate of the behavior of the true waveform.

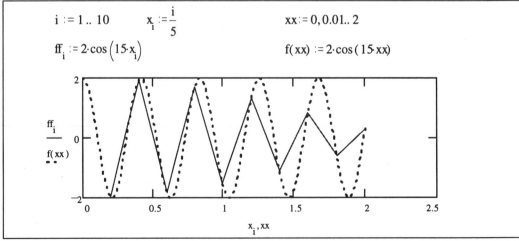

Figure 4

9.4 Fast Fourier Transform and Data File Structures

In the previous set of Explorations, we examined the synthesis and analysis of periodic wave-forms. We were able to not only construct waveforms from fundamental sinusoids and their harmonics but to break apart (supposedly) unknown waveforms into their discrete frequencies.

From a knowledge of the final product, we were able to determine the necessary components. And, from a knowledge of the signal's amplitude at each point within a sample, we were able to use averaging and summation techniques to estimate the harmonic content of the signal.

Here we will examine a more sophisticated tool, the Fast Fourier Transform, which will allow us to read data in from a structured data file (the sample of the unknown waveform), to then analyze the wave into its frequency spectrum and to finally filter out any noise.

As our sampled wave will be generated as a vector and noise added to it, we may compare the original signal content to the filtered wave as a check of the validity of the process.

Warmup

We will be analyzing a generated signal as if it were an unknown waveform, measured by a sampler and stored in a data file. The data file will contain information about the sampled amplitudes at specific indexed times along the signal's variation.

The created sample consists of a vector of 128 points generated from the sum of two sinusoids. The analyzed vector must have 2^m real values (with m, a positive integer) as a condition of the transform algorithm. Figure 1 shows the process of the data creation. You can apply this tool of analysis to any structured data file, any series of sampled measurements.

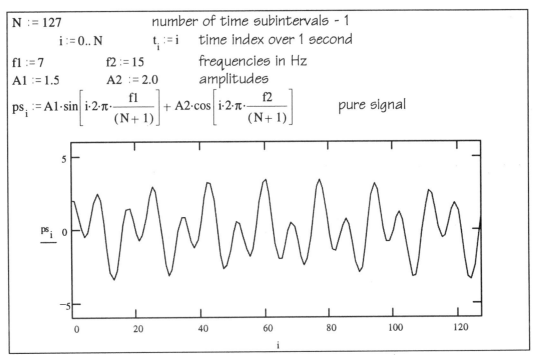

$N := 127$ number of time subintervals - 1

$i := 0..N$ $t_i := i$ time index over 1 second

$f1 := 7$ $f2 := 15$ frequencies in Hz

$A1 := 1.5$ $A2 := 2.0$ amplitudes

$$ps_i := A1 \cdot \sin\left[i \cdot 2 \cdot \pi \cdot \frac{f1}{(N+1)}\right] + A2 \cdot \cos\left[i \cdot 2 \cdot \pi \cdot \frac{f2}{(N+1)}\right] \qquad \text{pure signal}$$

Figure 1

At this point the sampled waveform and time intervals are stored in two vectors of 128 elements each. The total time interval is 1.0 second, broken down into 128 parts. Over this one second, the frequencies are 7 Hz and 15 Hz as defined in Figure 1.

The data pairs (time index, signal amplitude) need to be, for the purposes of using an unknown signal, written out to a data file that will retain the original formatting of the information, the connection between the time and the amplitude. For this reason, the data file, although simply a text file with numbers, needs to be structured.

Mathcad offers two options for the storage of data files. The first, WRITE(variable_name), simply writes a scalar to an unstructured text file. Numbers are stored in order of input with spaces between. The second, WRITEPRN(variable_name), offers more structure and is used for writing matrices to text files. Their complements are READ and READPRN, respectively.

☑ The READ-WRITE operations are covered in Chapter 5, Statistics, but are included here in case that chapter was omitted.

The sample information is stored in the two vectors, t_i and ps_i. In Figure 2, the WRITE operator writes the signal size (the number of samples) to a simple text file. The file 'psigsize.dat' is stored in the working directory you have specified for Mathcad. As a check, you can read the file in any text reader (e.g.: Windows's Notepad, DOS's Edit) and confirm that the number 128 has been stored there.

The actual sample data is stored using the operator WRITEPRN applied to the variable 'puresgnl'. The elements of the data file are defined as the augmentation of the t_i matrix with the ps_i matrix and are stored in the file 'puresgnl.prn'. Again, the file can be checked with a text editor.

☑ The augment(**x, y**) operator places vectors or arrays **x** and **y** side by side. The two separate vectors or arrays must have the same number of rows.

$\text{WRITE(psigsize)} := N + 1$	unstructured data file - sample size
$\text{PRNPRECISION} = 5$	set precision of stored data
$\text{WRITEPRN(puresgnl)} := \text{augment}(t, ps)$	sampled signal data

Figure 2

Now that the data has been successfully stored in a data file, we need to bring it back into our current document or a new document, to read it from the file, reconstruct it and compare it to the original. Figure 3 shows the process of reading the sample size and the structured data files. The data file is then broken apart into columns and the waveform is plotted.

☑ The columns of the matrix are defined using the matrix name followed by the column specification in superscript triangular brackets. This column operator is generated using CTRL-6. Attention should be paid to the index as the first physical column is defined as column number ZERO unless the ORIGIN variable has been reset to 1.

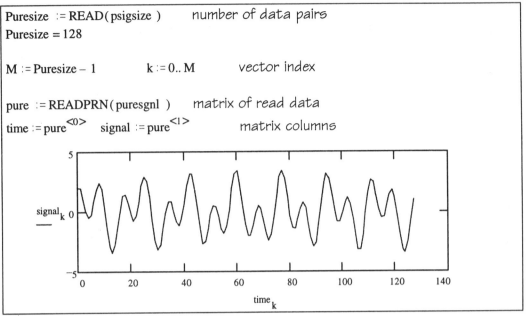

Puresize := READ(psigsize)   number of data pairs
Puresize = 128

M := Puresize − 1   k := 0.. M   vector index

pure := READPRN(puresgnl)   matrix of read data
time := pure$^{<0>}$ signal := pure$^{<1>}$ matrix columns

Figure 3

The reconstructed waveform has survived the WRITEPRN-READPRN cycle unscathed.

This process will now be applied to a signal containing noise. The signal will be generated and read into a matrix. This matrix will be analyzed for its frequency components using the Fast Fourier Transform utility in Mathcad. The utility (fft) is paired with its inverse (ifft) and operates on real data. A complex operator pair (cfft-icfft) is available for complex valued data.

The noisy signal is defined by setting a maximum noise level and adding a random value up to the maximum to the original pure signal, ps_i. An offset is added so that the random values are above and below the pure signal amplitudes. Figure 4 shows the definition and plot of the noisy signal and the writing of data to a structured text file.

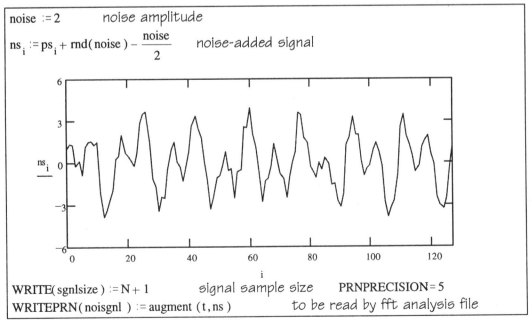

noise := 2   noise amplitude

$ns_i := ps_i + rnd(noise) - \dfrac{noise}{2}$ noise-added signal

WRITE(sgnlsize) := N + 1   signal sample size   PRNPRECISION = 5
WRITEPRN(noisgnl) := augment (t, ns)   to be read by fft analysis file

Figure 4

At this point, we have a data file that contains the yet-to-be-analyzed waveform. We will read the file into a new worksheet, plot the waveform then perform an analysis upon it. This analysis uses a more sophisticated method than the simple analysis of the coefficients covered in Section 9.3. A frequency spectrum will be generated and a noise reduction level set. By filtering out frequency information below this level, a purer wave can be reconstructed and the original frequencies regained.

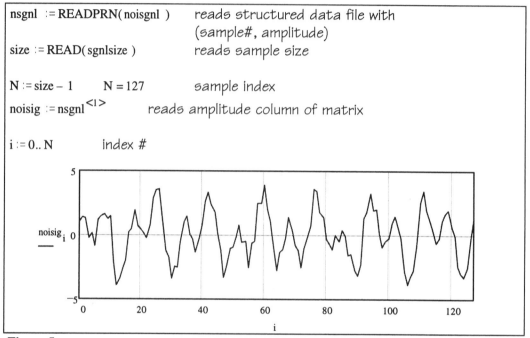

$\text{nsgnl} := \text{READPRN(noisgnl)}$ reads structured data file with (sample#, amplitude)

$\text{size} := \text{READ(sgnlsize)}$ reads sample size

$N := \text{size} - 1$ $N = 127$ sample index

$\text{noisig} := \text{nsgnl}^{<1>}$ reads amplitude column of matrix

$i := 0 .. N$ index #

Figure 5

As we had 128 samples in the original signal, the maximum frequency we can measure with any degree of confidence is 128/2 or 64 Hz. The fft operator is applied to the signal and a plot is made of the frequency versus the amplitude of the frequency component. These amplitudes correspond to the size of the coefficients in the previous method of analysis (Section 9.3).

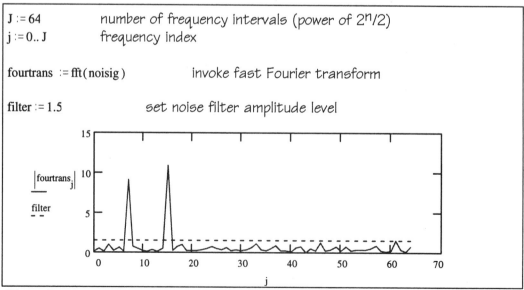

$J := 64$ number of frequency intervals (power of $2^n/2$)
$j := 0 .. J$ frequency index

$\text{fourtrans} := \text{fft(noisig)}$ invoke fast Fourier transform

$\text{filter} := 1.5$ set noise filter amplitude level

Figure 6

Note that the base frequencies of 7 Hz and 15 Hz generate clearly defined spikes in the frequency distribution while the added noise creates a low background of random frequency amplitudes.

☑ So as to wipe out the noise, the filter level was set to 1.0. Your results may differ depending on the randomness of your noise and the noise amplitude setting.

In order to extract the signal frequencies from the noise, the Heaviside step function was used. The function $\Phi(x)$ is equivalent to the 'if' operator applied as: if $(x < 0, 0, 1)$. The function $\Phi(x)$ returns a 1 if the argument is negative and a zero otherwise. The Greek letter Phi (upper case) is generated by typing F followed by (CTRL-g).

Figure 7 shows the filtering process and the reconstruction of the signal using the inverse fft operator.

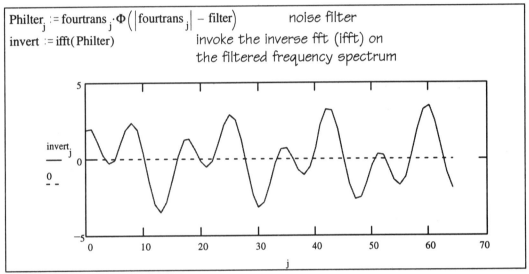

$$\text{Philter}_j := \text{fourtrans}_j \cdot \Phi\left(\left|\text{fourtrans}_j\right| - \text{filter}\right) \qquad \text{noise filter}$$

$$\text{invert} := \text{ifft}(\text{Philter})$$

invoke the inverse fft (ifft) on the filtered frequency spectrum

Figure 7

The indexed time information from $j = 65$ to $j = 127$ has been lost but that is a problem of the routine itself not of its application. The reconstructed wave is a fair copy of the original waveform.

In real applications, the strength of this Fourier Transform tool is the analysis of unknown-origin waveforms. This point will be covered in the Explorations which follow.

Explorations

The Basics

1. Use the READ/WRITE and READPRN/WRITEPRN utilities to store and read the data associated with the simple sinusoid $y(x) = 4 \sin(3x)$ over 3 periods. Create one file of data creation and another of data recovery. The number of vector elements is unimportant here as this sinusoid will not be fft analyzed.
2. Show that the wave in Figure 7 is less noisy than the waveform 'noisig' as a result of the filtering process. Analyze the filtered output and show that the original signal frequencies have been maintained while the overall noise level has been reduced.
3. The utility PRNPRECISION controls the precision of stored data in the structured files. Does the setting of the numerical precision control the sampling and filtering process in any way?

Beyond the Basics

1. Alter the level of noise in Figure 4 while keeping the noise filtering level (in Figure 6) a constant. At which point does the original signal get lost in the noise level?
2. Examine the effect of altering the noise filter level above and below the setting of filter = 1.5 on the reconstituted wave of Figure 7. If a high level of filtering is chosen, are there any negative effects on the remaining signal? What are the effects of too low a noise filtering level?
3. The fft/ifft algorithm requires that the number of elements within the analyzed vector be 2^m with m defined as a positive integer. Increase the sampling frequency by increasing the number of elements used in the exercise described in the Warmup. What are the effects of this increase on the filtering process? Is there a limit to the sampling frequency?
4. Create the messiest and noisiest waveform possible and save it to a structured data file. Exchange the data file for a similar one created by a classmate (for his/her waveform) and perform a Fourier analysis (with filtering) upon the unknown waveform. Compare your results to the definition of the original waveform as prescribed by your classmate. Be careful not to ruin your friendship.

Chapter 10: Linear Differential Equations

Newton and Leibniz did not create Calculus so that generations of students would have to survive a (mistakenly) feared course in Mathematics. You have already experienced their insight if you have taken the anti-derivative of a function. You have already solved a differential equation without being aware of it.

The derivative provides information about the system's change with time. It represents the dynamics of the system whether the system contains electrons, bacteria or money.

Most physical systems can be described by their dynamic equation, the equation which links the forces applied to them, the rates of change of their amplitudes, the rates of change of their rates of change, and so on...

These differential equations, containing derivatives or differentials, control the evolution of the system over time. The solution to these differential equations provides a time map of the system, an equation of growth (for bacteria and dollars) or motion (for massive objects). This map then becomes a powerful tool of prediction based on the system's initial conditions (initial amplitude, initial rate of growth, ...).

This chapter concentrates on numerical and graphical techniques applied to the solution of linear differential equations. As well, an especially powerful transformation tool, the Laplace Transform, will be introduced. This tool transforms the dynamic problem with all of its derivatives in time into an algebraic problem in frequency and is especially powerful for developing insight and efficiency in the solution of more complex differential equations.

10.1 Euler's Methods in Differential Equations

Numerous analytical approaches exist for solving the dynamic equations produced by physical systems. An exact solution is produced which satisfies the Differential Equation (DE) and the initial conditions imposed on it sometimes after considerable effort and loss of insight.

As the complexity of the DE increases, so does the length and technical difficulty of the analytic solution technique. To attend to this problem, the DE's have been broken down into classes and specific solution strategies have been developed for each of these.

In this set of exercises, you will explore an alternate approach to analysis. The use of numerical methods has already been examined in the cases of derivatives and integrals. Better and better estimates were available numerically as long as you were willing to wait for them. Here, the numerical process is applied to the solution of first-order Linear Differential Equations.

☑ The linearity of a Linear Differential Equation (LDE) refers to the way in which combinations of possible solutions are handled. If both $y_1(x)$ and $y_2(x)$ offer independent solutions to the LDE then so should the superposition $y_1(x) + y_2(x)$.

In most cases, exact solutions can be forced out of the DE. The application of a numerical approach may seem initially self-defeating. However, the study of non-linear forms would quickly reveal the limitations of the search for exact solutions. Rather, numerical methods can be used to approximate the solutions to the desired accuracy. These approximation techniques provide a rich ground for further study into the realms of non-linear DE's and fractals.

Warmup

In the case of a simple DE, $y'(x) = x^2$ with no set initial condition, the analytic solution is $y(x) = \dfrac{x^3}{3} + C$, with C a constant. The process of indefinite integration has been used to undo the differentiation defined by the DE.

A family of cubic curves is generated, each curve dependent on the to-be-defined initial state of the system. See Figure 1.

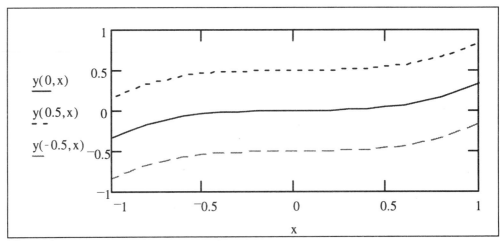

Figure 1

1. Direction Fields

A commonly used tool in the visualization of possible solution is the use of Direction (or Slope) Fields.

A direction field contains lines of unit or common length tangent at each grid point to the possible solution curves. Without an initial condition defined, this collection of lines covers the plane. The lines provide an overall sense of the solution to the DE since, at each point on the grid, the value of the first derivative defines a direction of increase or decrease for the function as the values of x and y are incremented. The second derivative indicates the local curvature.

Mathcad allows a similar approach in the application of its Vector Fields plot tool (available in Mathcad v6.0). A vector field includes a vector arrow (or directed line segment) at each grid point in the plane. The direction of the field is given by the direction of the arrow at that point while the amplitude is indicated by the relative length.

Figure 2 shows the vector field for the LDE mentioned above.

$k := 0 .. 10 \qquad j := 0 .. 10 \qquad$ indexes $\qquad dx := 0.2 \qquad$ stepsize

$x_k := -1 + k \cdot 0.2 \qquad y_j := -1 + j \cdot 0.2 \qquad$ grid $\qquad diff(x,y) := x^2 \qquad$ LDE

definitions of vector components...

$xcomp(x,y) := dx$

$ycomp(x,y) := if\left[diff(x,y) \leq 0, -\sqrt{(diff(x,y))^2 - dx^2}, \sqrt{(diff(x,y))^2 - dx^2} \right]$

definition of matrices of x and y components

$$X_{k,j} := xcomp\left(x_k, y_j\right) \qquad Y_{k,j} := ycomp\left(x_k, y_j\right)$$

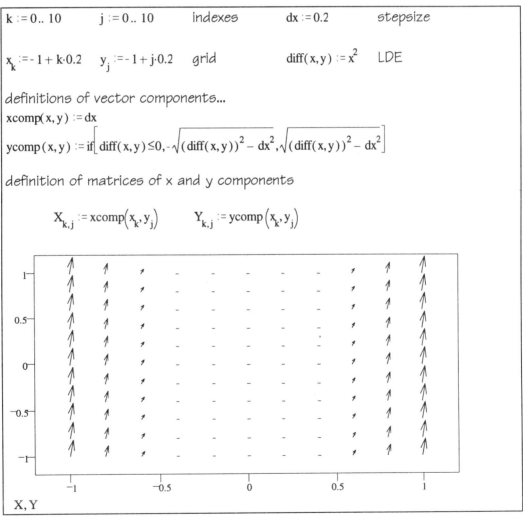

X, Y

Figure 2

In the region of the y-axis (where $x = 0$) the slope is minimal. As the absolute value of x grows, so does the rate of change of the function using the recipe prescribed by the LDE, $y'(x) = x^2$. As an example, at the point on the plane (0.5, 0.5) the derivative's value is 0.5^2 or 0.25. This is represented by an arrow of length 0.25 units pointed at an angle of $\tan^{-1}(.25)$ or 14.0°. As each point on the grid is mapped out in the same way, the overall direction of the family of solutions is revealed.

A comparison of the plot above with Figure 1 shows that the solution to the DE is within a family of cubic curves.

2. First-Order Euler Method

While the Slope Field approach is useful in determining the overall behavior of possible solutions, it does not provide an accurate representation of a particular solution for which the initial condition has been defined. Lines can be drawn between grid points based on the general direction of the slopes but these provide, at best, a rough estimate of the particular solution.

A more exact estimate of the solution uses the definition of the first derivative within the DE as the indicator of the function's first-order movement from point to point. Although crude, this method offers insight into the meaning of a derivative, the DE and its initial conditions.

For the differential equation, $y'(x) = f(x, y)$ with initial condition (x_0, y_0), the value of $f(x_0, y_0)$ at a point on the plane defines the immediate direction of the solution function $y(x)$ as x is incremented by an infinitesimal amount Δx away from x_0. The next point in the solution can be predicted to be $(x_1, y_1) = (x_0 + \Delta x, y_0 + y'(x_0) \cdot \Delta x)$.

At this point, $y'(x_1, y_1)$ will specify the new direction of the solution curve. This process is then repeated until a plot is obtained covering the desired domain in x. If the differential equation has an analytic solution, then a comparison can be made between the two complementary methods.

The process is repetitive and its accuracy dependent on the size of the increment Δx. Mathcad offers the use of vectors and the associated index for these types of problems. Figure 3 shows this particular method applied to a differential equation whose analytic solution has been worked out exactly (using the separation of variables technique) given the initial condition.

The function $f(x, y)$ can be defined with the x and y elements of a vector (x_i, y_i). The range of the index i depends on the range of x and the size of the increment Δx. Incremented values of x can then be assigned as $x_{i+1} := x_i + \Delta x$. The same indexing method can be applied to the y component so that $y_{i+1} = y_i + y'(x_i) \cdot \Delta x$ at each value of x_i.

In Figure 3, plot 3a shows the numerical solution overlaying the exact solution. There appears to be no difference in the plots. However, a closer inspection (plot 3b) reveals the two plots to be diverging from one another away from the seed point of the initial condition.

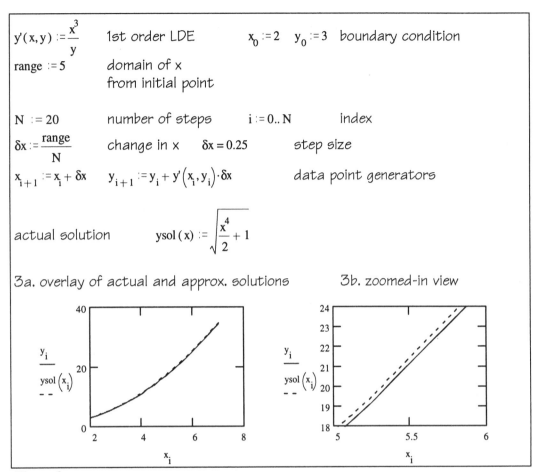

$$y'(x,y) := \frac{x^3}{y} \qquad \text{1st order LDE} \qquad x_0 := 2 \quad y_0 := 3 \quad \text{boundary condition}$$

range $:= 5$ domain of x
from initial point

$N := 20$ number of steps $i := 0..N$ index

$$\delta x := \frac{\text{range}}{N} \qquad \text{change in x} \quad \delta x = 0.25 \qquad \text{step size}$$

$$x_{i+1} := x_i + \delta x \qquad y_{i+1} := y_i + y'\left(x_i, y_i\right)\cdot\delta x \qquad \text{data point generators}$$

actual solution $ysol(x) := \sqrt{\dfrac{x^4}{2} + 1}$

3a. overlay of actual and approx. solutions   3b. zoomed-in view

Figure 3

The method applied to this LDE with a Δx of 0.25 provides a reasonable estimate of the solution. Figure 4 shows the associated error with respect to the exact solution.

In the case where an exact solution does not exist, a decision has to be made as to the acceptable degree of error. In the case of the examined DE, if error is desired of less than 1% then the Δx may have to be reduced and the number of points increased or an altogether better method examined.

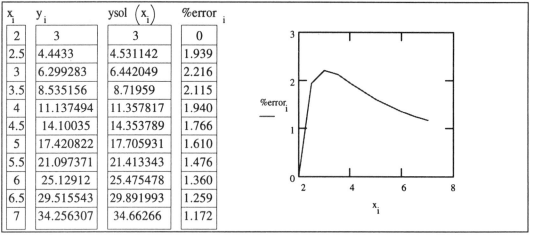

x_i	y_i	$ysol\left(x_i\right)$	$\%error_i$
2	3	3	0
2.5	4.4433	4.531142	1.939
3	6.299283	6.442049	2.216
3.5	8.535156	8.71959	2.115
4	11.137494	11.357817	1.940
4.5	14.10035	14.353789	1.766
5	17.420822	17.705931	1.610
5.5	21.097371	21.413343	1.476
6	25.12912	25.475478	1.360
6.5	29.515543	29.891993	1.259
7	34.256307	34.66266	1.172

Figure 4

3. Modified Euler's Method

In the previous method, the numerical solution has the tendency to wander away from the exact solution. A modification to the crudeness of this first method counters the wandering by predicting outwards and then by correcting for the error. This predictor-corrector method uses the average of the slopes evaluated at two points.

In Euler's method, the value of y_1 was predicted to be $y_0 + y'(x_0, y_0) \cdot \Delta x$ for a change in x_0 of Δx. The Modified Euler's Method attempts to correct any overshooting caused by the first order use of the differential Δx by also evaluating y' at (x_1, y_1) and then averaging these two y' results into a new and corrected value for y_1.

Finally, $y_1(corrected) = y_0 + \frac{1}{2}[y'(x_0, y_0) + y'(x_1, y_1)] \cdot \Delta x$

The process is then repeated over the whole domain of x at a step size of Δx. A comparison of the result of this method to the analytic solution should show a dramatic decrease in the error. Figure 5 shows the plots of the numerical and exact solutions for the DE and initial conditions from Figure 3. The function $ysol(x)$ indicates the exact solution to the DE.

☑ You may wonder at the appearance of the y_1 definitions in Figure 5 given that y_1 appears on both sides of the equation. In all programming languages, the $=$ sign does not imply equality but rather is interpreted as "the value of the right-hand side is *assigned* to the expression on the left-hand side". Thus a series of statements such as the following:

$$y_1 := y_0 + y'(x_0, y_0) \cdot \Delta x \qquad (1)$$

$$y_1 := y_0 + \frac{1}{2}[y'(x_0, y_0) + y'(x_1, y_1)] \cdot \Delta x \qquad (2)$$

is perfectly acceptable. In the second relation, the value of y_1 from (1) is used in the right-hand side. The new y_1 of relation (2) is then assigned the value of the result of the calculation.

$$x_{i+1} := x_i + \delta x$$

$$y_{i+1} := y_i + y'\left(x_i, y_i\right) \cdot \delta x \qquad \text{y value predictor}$$

$$y_{i+1} := y_i + \frac{y'\left(x_i, y_i\right) + y'\left(x_{i+1}, y_{i+1}\right)}{2} \cdot \delta x \qquad \text{y value corrector}$$

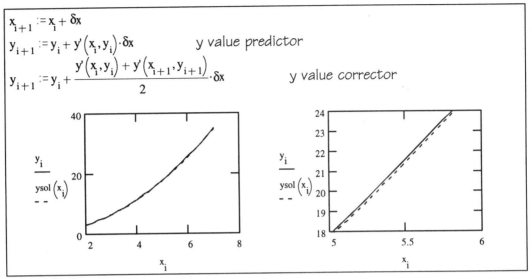

Figure 5

As expected, a dramatic improvement in accuracy has occurred. The plots are virtually indistinguishable. Figure 6 shows the error now below 1% over the indicated domain of the solution.

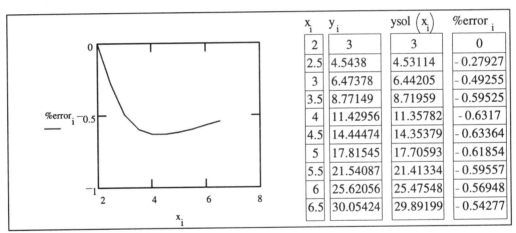

x_i	y_i	$ysol\left(x_i\right)$	$\%error_i$
2	3	3	0
2.5	4.5438	4.53114	-0.27927
3	6.47378	6.44205	-0.49255
3.5	8.77149	8.71959	-0.59525
4	11.42956	11.35782	-0.6317
4.5	14.44474	14.35379	-0.63364
5	17.81545	17.70593	-0.61854
5.5	21.54087	21.41334	-0.59557
6	25.62056	25.47548	-0.56948
6.5	30.05424	29.89199	-0.54277

Figure 6

4. Runge-Kutta Method

The last technique (named after German mathematicians Carl Runge and Wilhelm Kutta) of this set of exercises uses the weighted average of four prediction-correction slopes in order to obtain an extremely accurate picture of the behavior of the solution at a point.

For $x_1 = x_0 + \Delta x$, the slopes m in the vicinity of x_0 are given by

$$\text{1) } m_a = y'(x_0, y_0)$$
$$\text{2) } m_b = y'(x_0 + \Delta x/2, y_0 + m_a \cdot \Delta x/2)$$
$$\text{3) } m_c = y'(x_0 + \Delta x/2, y_0 + m_b \cdot \Delta x/2)$$
$$\text{4) } m_d = y'(x_0 + \Delta x, y_0 + m_c \cdot \Delta x)$$

The weighted average of these slopes is $m_{avg}(x_0, y_0) = \frac{1}{6}[m_a + 2 \cdot m_b + 2 \cdot m_c + m_d]$.

The predicted and corrected value of y_1 is $y_1 = y_0 + m_{avg}(x_0, y_0) \cdot \Delta x$.

This process is then repeated as the index is incremented over the whole range. Figure 7 shows the application of this last technique to our original DE.

$$x_{i+1} := x_i + \delta x \qquad \text{with} \qquad \delta x = 0.25 \qquad \text{increase in x by stepsize}$$

y value predictors using average of
four slopes in vicinity of x_i

$$m_a(x, y) := dy(x, y) \qquad\qquad m_b(x, y) := dy\left(x + \frac{\delta x}{2}, y + m_a(x, y) \cdot \frac{\delta x}{2}\right)$$

$$m_c(x, y) := dy\left(x + \frac{\delta x}{2}, y + m_{.b}(x, y) \cdot \frac{\delta x}{2}\right) \qquad m_d(x, y) := dy\left(x + \delta x, y + m_c(x, y) \cdot \delta x\right)$$

m_{avg} is the weighted average of the slopes m_a, m_b, m_c, m_d used to
predict the approximate location of y_{i+1}

$$m_{avg}(x, y) := \frac{1}{6} \cdot \left(m_a(x, y) + 2 \cdot m_{.b}(x, y) + 2 \cdot m_c(x, y) + m_d(x, y)\right)$$

$$y_{i+1} := y_i + m_{avg}(x_i, y_i) \cdot \delta x \qquad\qquad \text{y value corrector}$$

Figure 7

The error associated with the numerical process (Figure 8) has virtually disappeared over the domain of *x*.

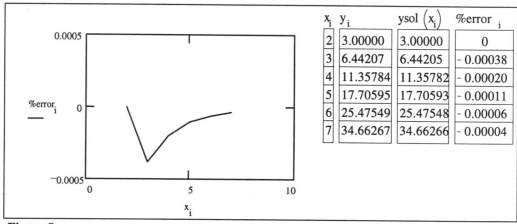

x_i	y_i	ysol$\left(x_i\right)$	%error$_i$
2	3.00000	3.00000	0
3	6.44207	6.44205	- 0.00038
4	11.35784	11.35782	- 0.00020
5	17.70595	17.70593	- 0.00011
6	25.47549	25.47548	- 0.00006
7	34.66267	34.66266	- 0.00004

Figure 8

Explorations

The Basics

1. Apply each of the three techniques (Euler, Modified Euler, Runge-Kutta) to the DE, $y' + x = 2$ with the initial condition (4, 0). The initial condition provides a seed point from which to start the numerical analysis. Choose an appropriate step-size and indicate the error between the numerical and exact solutions with the use of tables and plots.

2. Apply the three techniques to the DE, $\dfrac{dy}{dx} + yx = 0$ for the initial condition (0.0, 1.5).

3. A physical system (electrons, bacteria, or money) has a growth rate dependent on $\dfrac{dP}{dt} = 0.4P - 200$. Using one of the three numerical methods, determine the behavior of the system for each of the three following initial conditions:
   a) $P = 1000$ at $t = 0$ s
   b) $P = 500$ at $t = 0$ s
   c) $P = 200$ at $t = 0$ s.

Beyond the Basics

1. For the DE of *The Basics* #1, examine the effect of decreasing the step-size Δx within the Euler method. How do you decide at which resolution to stop considering that infinite accuracy would require infinite time? At which point is the error comparable to that generated by the Runge-Kutta method applied at the original step-size.

2. Compare, in terms of accuracy and speed of calculation, the application of each of the three methods to the DE, $y' = \dfrac{\ln(x)}{x^2 + y}$ for the initial condition (2.0, 5.0). How is the domain of x restricted?

3. (Version 6.0) Use the prescription in Figure 2 to create a vector field for the DE, $y' = \dfrac{x^3}{y}$.

10.2 The Laplace Transform

Examination of the analytic techniques applied to differential equations shows the overwhelming need for classification. Pencil and paper solutions can span many pages. And, in fact, the differential equations do require this depth of analysis. The mutual dependence of the various orders of derivatives necessarily leads to lengthy analysis.

As with other areas of mathematics and technology (and life), problems can initially seem difficult or even impossible to solve. If total avoidance of the problem is not an option, the translation of the problem into another language or frame of reference can allow an easier solution path. We use this tool in everyday conversation when we try to explain a situation to someone who refuses to understand our clear reasoning. We say, "It's like a ... (fill in the appropriate simile/comparison)." Solution by metaphor! The other person answers, "Oh, of course! I understand now." And, indeed, understanding has been achieved.

The difficulty with and strength of differential equations is the necessary interdependence of the various orders of the rate of change. The many variations these relations can have leads to their classification into families. Specific analytic strategies can then be applied to the solution process. Some forms, however, do not allow closed-form solutions. Then, the numerical techniques of Section 9.1 can be applied.

The Laplace Transform (P.S. Laplace, France, 1749-1827) is a process of translating the rate-of-change elements of the differential equations into pure algebraic elements. The function and its derivatives change into algebraic expressions complete with initial conditions. The solution of the translated problem is the solution of an algebraic equation and yet, in all ways, is the solution to the original DE. None of the original information is lost. And, when the translation is inverted, the solution is a perfectly accurate picture of the response of the system.

This circular process is used in many solution strategies and is defined in Figure 1.

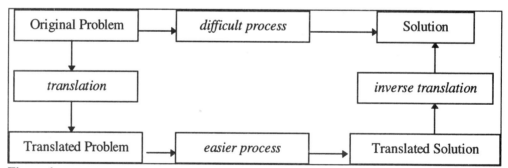

Figure 1

The Laplace Transform offers a translation strategy which turns the problem in rates of change into a problem of algebraic manipulation. The translated problem should allow a quicker and easier solution. The process of translation is then inverted so that the solution is understandable within the original language.

If the original DE is in the time domain (i.e., contains derivatives with respect to time), the translation process takes the problem into the frequency domain. The frequency response of the system is made clear by the form of the translated solution. At this point, you can choose to either remain in the frequency domain or translate back to the time domain.

Warmup

The Laplace Transform of a function $f(t)$ is given by $L[f(t)] = \int_0^{+\infty} f(t)e^{-st}\,dt$. The Transform is a function of the complex variable $s = \sigma + j\omega$ and is often expressed as $F(s)$.

☑ The Laplace Transform utility is only available in Version 6.0 and is invoked by selecting Symbolic/Transforms/Laplace Transform. The Symbolic/Evaluate Symbolically utility can be used in Version 5.0 to evaluate the Laplace integral for a function $f(t)$. However, exact forms may not exist for many simple functions. Both versions offer a Symbolic/Convert to Partial Fractions utility that is used to simplify the Laplace expression before inversion.

Figure 2 includes the application of the Laplace Transform to some simple functions. By inspection, you can see that for functions of the form $f(t) = t^n$, $F(s) = \dfrac{n!}{s^{n+1}}$ for $n = 0, 1, 2...$

functions f(t) =				
(a)	(b)	(c)	(d)	(e)
1	t	t^2	t^3	t^9
have Laplace transforms...				
$\dfrac{1}{s}$	$\dfrac{1}{s^2}$	$\dfrac{2}{s^3}$	$\dfrac{6}{s^4}$	$\dfrac{362880}{s^{10}}$

Figure 2

Figure 3 contains the Laplace Transforms of some transcendental functions.

functions f(t) =			
(a)	(b)	(c)	(d)
$e^{a\cdot t}$	$\cos(a\cdot t)$	$\sin(a\cdot t)$	$t\cdot e^{a\cdot t}$
have Laplace transforms...			
$\dfrac{1}{(s-a)}$	$\dfrac{s}{\left(s^2+a^2\right)}$	$\dfrac{a}{\left(s^2+a^2\right)}$	$\dfrac{1}{(s-a)^2}$

Figure 3

The Laplace Transform is a linear process and so displays the properties associated with linear operations. Figure 4 displays the following properties of the transform:
   a) constant times a function
   b) sum of functions
   c) general function
   d) derivative of a general function
   e) derivative of an defined function outside the derivative operator
   f) derivative of a defined function
   g) integral of a general function
   h) integral of a defined function
Notice that for expressions of generalized functions $f(t)$, the transform process returns a generalized form (cases (c), (d) and (g)).

As applied to the derivative of a function defined outside the derivative operator (case (e)), the process returns a generalized form of the rule $L[y'(t)] = s \cdot L[y(t)] - y(0)$. There is no need for a constant $y(0)$ in case (f) as the derivative outputs $12 \cdot t^3$ to which the Transform is applied.

The transform of the integral of a general function in case (g) agrees with the analytical result $F(s)/s$. Case (h) illustrates the transform of an integral of a specific function $f(t) = t$.

☑ In most texts, the integral of the function $f(t)$ to be transformed is represented

as $\int_0^t f(t)dt$. This representation may not cause confusion in a textbook but it most certainly would confuse any software program. Here, in cases (g) and (h), we have defined the Laplace Transform of the integral of a function as

$L[f(T)] = L[\int_0^T f(t)dt]$. The integral in $f(t)$ produces a secondary function in T to which the Laplace Transform is then applied. As an example, in case (h), the general analytic result would produce $\frac{1}{s} \cdot L[t] = \frac{1}{s} \cdot \frac{1}{s^2}$. As Mathcad processes

the same integral, the Laplace Transform is applied to $\frac{T^2}{2}$ the output of the

integral. This produces $\frac{1}{2} \cdot \frac{2!}{s^3}$, the same result.

(a)	(b)	(c)	(d)
$9 \cdot t^2$	$t + t^3$	$f(t)$	$\frac{d}{dt} f(t)$
has Laplace transform	has Laplace transform	has Laplace transform	has Laplace transform
$\frac{18}{s^3}$	$\frac{1}{s^2} + \frac{6}{s^4}$	$\text{laplace}(f(t),t,s)$	$\text{laplace}(f(t),t,s) \cdot s - f(0)$

(e)	(f)	(g)	(h)
for $f(t) = 3 \cdot t^4$ $\frac{d}{dt} f(t)$	$\frac{d}{dt} 3 \cdot t^4$	$\int_0^T f(t)\, dt$	$\int_0^T t\, dt$
has Laplace transform	has Laplace transform	has Laplace transform	has Laplace transform
$\text{laplace}(f(t),t,s) \cdot s - f(0)$	$\frac{72}{s^4}$	$\frac{1}{s} \cdot \text{laplace}(f(T),T,s)$	$\frac{1}{s^3}$

Figure 4

All well and good. But we have not yet applied these tools to the translation and solution of a differential equation. For the first-order differential equation $\frac{dy}{dt} = t$ with $y(0) = 5$, we can define the differential equation using the CTRL = sign for the symbolic calculation. However, as given in Figure 5, only a generalized version of the translated problem emerges from the application of the Laplace Transform utility.

$$\frac{d}{dt}y(t)\mathbf{=}t \qquad \text{(then select t) has transform} \qquad laplace(y(t),t,s)\cdot s - y(0)\mathbf{=}\frac{1}{s^2}$$

Figure 5

The translated equation was then edited, the laplace(*w*(*t*), *t*, *s*) expression replaced with the variable *L* and the *y*(0) initial condition replaced with the number 5.

☑ This editing process probably seems like a tedious exercise for this level of simple DE. As the complexity of the equations grows, the value of the translation tool will become more evident.

Figure 6 shows the edited version of the translated differential equation. The variable *L* is the Laplace translation of the solution. Since the form of the equation is pure algebraic, this variable can be isolated by using Symbolic/Solve for Variable. Select *L* by clicking beside it and using the Up-arrow once. Then, isolate *L*.

$$laplace(y(t),t,s)\cdot s - y(0)\mathbf{=}\frac{1}{s^2} \qquad \text{replace laplace by L and}$$

$$\text{y(0) by value then isolate}$$

$$L\cdot s - 5\mathbf{=}\frac{1}{s^2} \qquad \text{has solution(s) for L} \qquad \frac{-\left(-5-\frac{1}{s^2}\right)}{s}$$

Figure 6

At this point, you have solved the problem. However, the solution is still in the translated Laplace-*s* language and must be translated back to the time-*t* language. After selecting the variable *s* within the expression for *L*, select Symbolic/Transforms/Inverse Laplace Transform. Figure 7 shows the output for the process. The expression was then copied into a definition of the solution function *y*(*t*) and plotted.

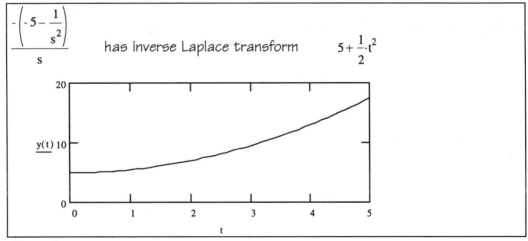

$$\frac{-\left(-5-\frac{1}{s^2}\right)}{s} \qquad \text{has inverse Laplace transform} \qquad 5+\frac{1}{2}\cdot t^2$$

Figure 7

This process can be applied as easily to a second-order differential equation. However, the process of translation is not so immediate. For the equation,

$$\frac{d^2}{dt^2} y(t) + 5 \cdot \frac{d}{dt} y(t) + 5 \cdot y(t) = t$$

with $y(0) = 2$ and $y'(0) = 2$, the Laplace Transform in Mathcad returns an accurate but not immediately decipherable expression. However, selection of individual components reveals the structure of the transforms. The output for the second derivative still results in a message that the transform is peculiar to the symbolic process and has been stored in the Clipboard. This is equivalent to the first part of the expression given in Figure 8 and represents $L[y''(t)] = s^2 L[y(t)] - s \cdot y(0) - y'(0)$.

$$\frac{d^2}{dt^2} y(t) + 5 \cdot \frac{d}{dt} y(t) + 5 \cdot y(t) \equiv t_0 \quad \ldots \text{2nd-order differential equation}$$

Clipboard translation...
(((((laplace(y(t),t,s))*(s))+((-1)*(y(0))))*(s))+
((-1)*(diff(y(t2),t2)))+((5)*(laplace(y(t),t,s))*(s))+
((-5)*(y(0)))+((5)*(laplace(y(t),t,s))))=((s)^(-2)) &where {(t2)=(0)}

Individual term translation

$$\frac{d^2}{dt^2} y(t) \qquad \text{result saved in clipboard}$$

$5 \cdot \frac{d}{dt} y(t)$	t	$5 \cdot y(t)$
has Laplace transform	has Laplace transform	has Laplace transform
$5 \cdot \text{laplace}(y(t),t,s) \cdot s - 5 \cdot y(0)$	$5 \cdot \text{laplace}(y(t),t,s)$	$\dfrac{1}{s^2}$

Figure 8

The translated equation then needs to be expressed in terms of L and the initial conditions for the function $y(t)$ and its first derivative. The variable L is then isolated and the Inverse Laplace Transform is taken. As illustrated in Figure 9, the solution does not present an immediately visible behavior. The output can be easily copied and pasted into the definition of the solution function and a plot can be drawn over a suitable range.

Note that the time behavior of the solution seems to be broken down into two sections. The first, from $t = 0$ seconds to approximately 2.0 seconds, reveals a short-run (or transient) behavior largely controlled by the exponential terms. The second section, for $t > 2.0$ seconds, shows the long-run behavior of the function controlled by the linear term.

☑ The terms resembling $\cosh(x)$ and $\sinh(x)$ in Figure 9 are the hyperbolic cosine and hyperbolic sine, respectively. They are defined as:

$$\cosh(x) = \frac{e^x + e^{-x}}{2} \quad \text{and} \quad \sinh(x) = \frac{e^x - e^{-x}}{2}.$$

translated equation with y(0)=2 and y'(0)=2

$$L \cdot s^2 - s \cdot 2 - 2 + 5 \cdot s \cdot L - 5 \cdot 2 + 5 \cdot L = \frac{1}{s^2}$$ transformed equation

isolated expression for L[y(t)] $\dfrac{-\left(-2 \cdot s - 12 - \dfrac{1}{s^2}\right)}{\left(s^2 + 5 \cdot s + 5\right)}$

inverse transform

$$\frac{1}{5} \cdot t - \frac{1}{5} + \frac{73}{25} \cdot \exp\left(\frac{-5}{2} \cdot t\right) \cdot \sinh\left(\frac{1}{2} \cdot \sqrt{5} \cdot t\right) \cdot \sqrt{5} + \frac{11}{5} \cdot \exp\left(\frac{-5}{2} \cdot t\right) \cdot \cosh\left(\frac{1}{2} \cdot \sqrt{5} \cdot t\right)$$

$t := 0, 0.1 .. 10$

$$y(t) := \frac{1}{5} \cdot t - \frac{1}{5} + \frac{73}{25} \cdot \exp\left(\frac{-5}{2} \cdot t\right) \cdot \sinh\left(\frac{1}{2} \cdot \sqrt{5} \cdot t\right) \cdot \sqrt{5} + \frac{11}{5} \cdot \exp\left(\frac{-5}{2} \cdot t\right) \cdot \cosh\left(\frac{1}{2} \cdot \sqrt{5} \cdot t\right)$$

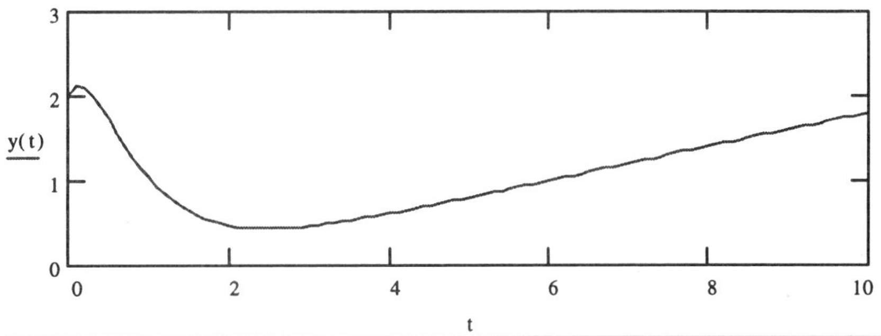

Figure 9

An Electronic Application

In electronic circuits, the closing of a switch is viewed as the point at which the circuit becomes active. All the circuit elements respond to the forcing function of the voltage supply, either a DC or AC source.

As an example of both the power of differential equations and the Laplace Transform, we will examine a simple circuit with a resistor (*R*) and inductor (*L*) in series with an alternating current generator (*v*). The elements form a closed loop controlled by the opening and closing of a switch.

Although the resistor acts as a pure resistive device, the inductor reacts to the constant changes imposed upon it by the alternating forcing function. This causes a shift in the phase of the voltage across the inductor relative to the voltage across the resistor.

The overall response of the circuit contains a transient part, a response to the switch being turned on at *t* = 0, and a steady state response which describes the system's behavior after the transient has exhausted itself.

The method used in the analysis is exactly that developed within the Warmup.

The differential equation generated for a circuit with $R = 2000$ Ohms, $L = 200$ mHenries and $v(t) = 50 \cos (2\pi f t)$ with $f = 600$ Hz is given by :

a) in general, $R \cdot i(t) + L \cdot \dfrac{di(t)}{dt} = v \cdot \cos(\omega \cdot t)$ and,

b) in this particular case, $2000 \cdot i(t) + 0.200 \cdot \dfrac{di(t)}{dt} = 50 \cdot \cos(1200\pi \cdot t)$

The initial condition is $i(0) = 2$ Amps. Figure 10 shows the solution process with the inductance defined as *LL* so as not to confuse it with the Laplace Transform, *L*.

☑ The output of the inverse transform process created an expression with a numerical precision of 20 (the symbolic default value). Although precise, these numbers are not easily absorbed. From the Symbolic menu, the Evaluate/Floating Point option was used, set to a precision of 3.

☑ The inverse transform answer extended beyond the range of the page and was herded back using the 'Break with Plus' operation. Select the left hand expression up to and including the term you would like the break to appear before. Press Delete to delete the operator before this last term. Then press CTRL-Enter to create the wrap-around expression.

Series RL circuit with AC source

$R := 2000 \, \Omega$ $LL := 0.200 \, \text{henry}$ ...variables, reference

$\dfrac{R}{LL} = 1 \cdot 10^4 \cdot \sec^{-1}$ $f := 600 \, \text{Hz}$ $\omega := 2 \cdot \pi \cdot f$ $\omega = 3.77 \, 10^3 \cdot \sec^{-1}$

$R \cdot i(t) + L \dfrac{d}{dt} i(t) = v \cdot \cos(\omega \cdot t)$ ... general equation

$2000 \, i(t) + 0.200 \dfrac{d}{dt} i(t) = 50 \cos(1200\pi \cdot t)$ ...specific equation

(1) has Laplace Transform--after editing and with i(0) = 2 --of

$2000 \cdot L + .2 \cdot L \cdot s - .2 \cdot 2 = 50 \cdot \dfrac{s}{\left(s^2 + 14212230.3375686764\right)}$

(2) select Symbolic/Evaluate/Floating Point to simplify

$2000 \cdot L + .2 \cdot L \cdot s - .2 \cdot 2 = 50 \cdot \dfrac{s}{\left(s^2 + 1.42 \, 10^7\right)}$

(3) isolate L $-1 \cdot \dfrac{\left[-.4 - 50 \cdot \dfrac{s}{\left(s^2 + 14200000\right)} \right]}{(2000. + .2 \cdot s)}$

(4) take Inverse Transform and use Break with Plus

$\Bigg(1.9781085814360770578 \cdot \exp(-10000. \cdot t) \, ... $

$+ 8.2493185995695148616 \cdot 10^{-3} \cdot \sin(3768.2887362833543888 \cdot t) \Bigg) \, ...$

$+ 2.1891418563922942207 \cdot 10^{-2} \cdot \cos(3768.2887362833543888 \cdot t)$

Figure 10

This precise output was edited to a more reasonable 3 digit precision and defined as $i(t)$. Figure 11 shows the current function up to 40 ms and indicates a transition point from the transient to steady-state response at 5 times the value of the ratio $(LL)/R$ (the time constant).

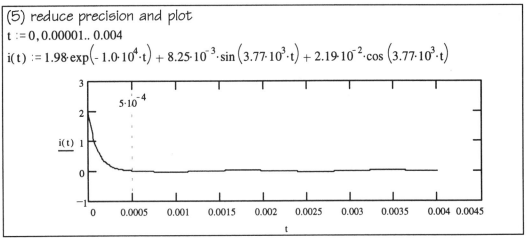

(5) reduce precision and plot

$t := 0, 0.00001.. 0.004$

$i(t) := 1.98 \cdot \exp\left(-1.0 \cdot 10^4 \cdot t\right) + 8.25 \cdot 10^{-3} \cdot \sin\left(3.77 \cdot 10^3 \cdot t\right) + 2.19 \cdot 10^{-2} \cdot \cos\left(3.77 \cdot 10^3 \cdot t\right)$

Figure 11

The response to the AC forcing function seems to be an exponential decay from the initial condition to a constant level. However, if we expand the range of the 'constant' level, the true behavior of the circuit is revealed. Figure 12 shows an expanded view of i (t) with a superimposed scaled view of the forcing function variation.

☑ The scaled view is necessary as we are using the same y-axis to represent both voltage and current.

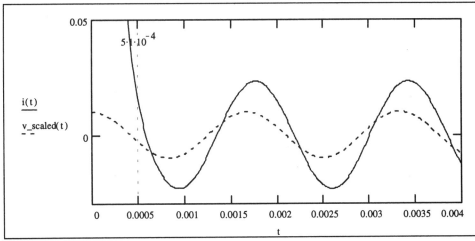

Figure 12

The current's steady-state frequency matches that of the voltage. However, the sinusoids are slightly out of phase, the time difference of approximately 0.1 ms an indication of the reactance of the inductor to the repeated pushes and pulls of the voltage.

Explorations

The Basics

1. Solve the first order differential equation, $y'(t) + 2y = 3e^{-2t}$, with initial condition $y(0) = 0.5$ using the technique developed in the Warmup.
2. Solve the second-order differential equation, $4y'' + 6y' - 3y = 0$, with initial conditions $y(0) = 1$ and $y'(0) = 1$.
3. Examine the behavior of a simple series RL circuit with a stable DC source. The switch is closed at $t = 0$ at which time the current in the circuit is given by $i(0) = 0$. The value for the resistance is $R = 2000$ Ohms while the inductor's size is 0.200 Henries and the voltage is 50.0 volts.
4. If the inductor in the previous exercise is replaced with a Capacitor, the respective expression in the differential equation is replaced by $\dfrac{1}{C} \cdot \displaystyle\int_0^t i(t)dt$ which has a Laplace Transform of $\dfrac{1}{C} \cdot \dfrac{L[i(t)]}{s}$. Repeat the analysis procedure for a capacitor of 500 µF. How does the circuit respond for this type of reactive component?

Beyond the Basics

1. Repeat the analysis of Figures 10, 11 and 12 for a series of resistive values from 500 Ohms to 3000 Ohms in steps of 500 (e.g.: trial 1 with $R = 500\ \Omega$, trial 2 with $R = 1000\ \Omega$, and so on). Can you determine the effect of the resistance on the response of the circuit?
2. With reference to *Beyond the Basics* #1, analyze the effect of various inductor sizes on the response of the circuit.
3. For the second-order differential equation of Figures 8 and 9, examine the effect on the response of changing the values of the coefficients of $y(t)$ and $y'(t)$. In second-order equations, why would you expect to see functions of sine, cosine and the exponential repeatedly show up in the solutions?
4. A series LRC circuit is an electrical circuit with a resistor (R), capacitor (C) and inductor (L) in series with a voltage source. For a DC source turned on at $t = 0$ s, the differential equation is $R \cdot i(t) + L \cdot \dfrac{di(t)}{dt} + \dfrac{1}{C} \displaystyle\int_0^t i(t)dt = V$. The general transform of the equation is $R \cdot L + LL \cdot (L \cdot s - i(0)) + \dfrac{1}{C} \cdot \dfrac{L}{s} = \dfrac{V}{s}$. With initial conditions of $i(0) = 0$ and $i'(0) = 0$ and for $R = 2000\ \Omega$, $L = 0.200$ H, $C = 500$ µF and $V = 50$ volts, determine the response of the system. These particular values create a condition known as overdamping where $\dfrac{R}{2L} > \sqrt{\dfrac{1}{LC}}$.
5. Determine the response of the system outlined in *Beyond the Basics* #4, for values of R, L and C which satisfy $\dfrac{R}{2L} < \sqrt{\dfrac{1}{LC}}$. This condition is recognized as underdamping.
6. Repeat the process of #4 for the critically damped condition, $\dfrac{R}{2L} = \sqrt{\dfrac{1}{LC}}$.

Appendices

These additional sections are meant as references for those students wishing to learn more about some of the topics or utilities used within the rest of the text. There are no Explorations as such although Exercises are included to illustrate the ideas.

1 Coordinate Axis Rotation
2 Linear Regression Theory

1. Coordinate Axis Rotation

In Chapter 3, Section 8, on Parametric Curves and Polar Coordinates, some of the functions you examined were symmetric with respect to the origin. These functions were expressed in polar rather than rectangular coordinates. The strength of one form of expression versus the other depends on the particular relation and its visual symmetry. Although you can express a linear relation in polar form, the effort required defeats itself. The same applies to expressing complex polar forms in rectangular coordinates.

One of the explorations within Section 3.8 examined the process of rotating any of the relations by a fixed angle about the origin. The definition of a set of axes is largely arbitrary, a matter of reference frame or preference of origin. A set of points in one reference frame may not have the same coordinates in another rotated with respect to the first.

The same relativity applies to origins which have been translated.

Theory

The rotation of a point by an angle θ around the origin of a fixed coordinate system creates a final position which is impossible to differentiate from that of the point having remained fixed and the coordinate axes having rotated by -θ.

The rotation of coordinate axes assumes the point to be described is fixed (at least at an instant of time) and that the coordinate axes rotate with respect to one another about the common origin. The original or rotated coordinates then describe the same point in terms of its coordinates relative to the orientation of the reference frame.

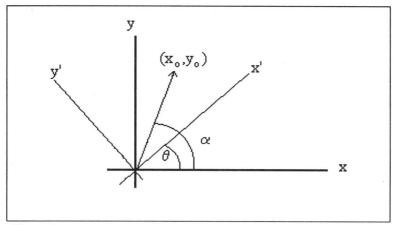

Figure 1

An analysis of Figure 1 shows that a point (x_0, y_0) in the original two-dimensional x-y coordinate system can be represented by its polar form $(r \cos\alpha, r \sin\alpha)$ where r is the radius and α is the angle the radius makes with the positive x-axis. In the frame x'-y' rotated an angle θ with respect to the positive x-axis, the coordinates of the same (fixed) point would be represented by:

$$x_0' = r \cos(\alpha - \theta)$$
$$y_0' = r \sin(\alpha - \theta)$$

From the trigonometric identities for the sum and difference of angles, these relations transform into:

$$x_0' = r\cos(\alpha)\cos(\theta) + r\sin(\alpha)\sin(\theta)$$
$$y_0' = r\sin(\alpha)\cos(\theta) - r\cos(\alpha)\sin(\theta)$$

With our previous polar definitions of x_0 and y_0, these relations can be rewritten as:

$$x_0' = x_0\cos(\theta) + y_0\sin(\theta)$$
$$y_0' = -x_0\sin(\theta) + y_0\cos(\theta)$$

A more compact and elegant way of representing this coordinate transformation between rotated axes is the use of a rotation matrix. Thus,

$$\begin{pmatrix} x_0' \\ y_0' \end{pmatrix} = \begin{pmatrix} \cos\theta & \sin\theta \\ -\sin\theta & \cos\theta \end{pmatrix} \begin{pmatrix} x_0 \\ y_0 \end{pmatrix}$$

This matrix produces the new coordinates of the point as measured in the rotated reference frame.

Exercises

1. In the above generalized rotation matrix, the determinant is 1 for any value of the angle θ. What is the significance of this (if any)?
2. Rotate the coordinate point (-4.5, 6.2) by +30° about the origin. Now, rotate the same point by -330°. Is there any difference between the two results? How is the rotation of the coordinate point different from the rotation of the coordinate axes?
3. Create a circle of radius 4 centered at the point (6, 5). Rotate the coordinate axes by -25°. Where is the center of the circle located now?
4. Create an ellipse, centered at the origin, with a semi-major axis of 6 units and a semi-minor axis of 3 units. How would you rotate this ellipse by +45° with respect to the origin?
5. How would you define a circle centered at the origin in terms of the effect of coordinate axis rotation upon it?
6. What changes would you expect in the 2-dimensional rotation matrix if it were generalized to 3 dimensions? Give an example of the new transformation in θ and ϕ as applied to a 3-dimensional vector (x, y, z).

2. Linear Regression Theory

Within Chapter 5 on Statistics, utilities such as **linterp**, **slope**, **intercept** and **corr** were used to find interpolated data and, more importantly, the best-fit line through your data. Mathcad performs these analytical functions under the page, in a set of subroutines called by the respective names.

The linear regression work you have done can be completed at the level of its definition with even the simplest calculator. However, the use of summation symbols and the ability to automate processes allows a mid-point between the obscurity of simply calling a routine to do your work and the other (mind-numbing) extreme of performing all calculations with pencil and paper.

The data set you have examined is made up of coordinate points each of which exhibits a reading or systematic error. There is an essential indistinctness to data points. Each point is, to some degree, unknown. Here, however, we are looking at the degree to which seemingly linear data may diverge from the perfect line. By balancing the divergences, a line which best fits the data can be constructed and measures taken of its reliability. We have assumed that the data is pure and that your expectations of its behavior have not altered your analysis of its behavior.

Theory

You have performed measurements on a system and have collected N data pairs which you suspect have a linear relation. You may have plotted the points already and this has given you the clue. Regardless, you can analyze them for their inherent or apparent linearity.

For the N pairs of points (x_i, y_i) where i is a numbering index from 1 to N (or from 0 to N-1), you assume your points fit the straight line function $y = mx + b$ where m is the slope of the line and b is the y-axis intercept. The point of any analysis is then to determine the best slope and intercept for your data. Often this is done in the lab using a transparent ruler and eyeballing the line. Although quick and dirty, this method suffers from just a bit of inaccuracy.

If you assume a best-fit line exists and is defined by slope m and intercept b, then the deviation of your data point from its theoretical position is given by $D_i = y_i - (mx_i + b)$. This difference represents the vertical distance between your data point and the line. For some of your points, the deviation will be positive; for other, negative. So that distances are measured rather than displacements, the square of this deviation is taken. Then, for an individual point,

$$D_i^2 = (y_i - mx_i - b)^2 .$$

The sum of the squared deviation for all N points is then given by

$$D^2 = \sum_{i=1}^{N} D_i^2$$

$$= \sum_{i=1}^{N} (y_i - mx_i - b)^2 .$$

The definition of the line of best-fit is the line for which this expression is a minimum, the line which has the "least squares". Different choices of m and b will create different total deviations. The squared deviations do not reduce to zero unless the data is absolutely and perfectly linear (any real data that satisfied this condition would be suspect). The process of "least squares" is an attempt to minimize the overall deviation and find the best fit of many possibilities.

As D^2 is a function of the two variables m and b, its minimum occurs where the partial derivatives (i.e., a derivative with respect to one variable while treating the other variable as constant) equal zero.

$$\frac{\partial D^2}{\partial m} = 0 \text{ and } \frac{\partial D^2}{\partial b} = 0$$

These two conditions applied to the definition of D^2 yield the following relationships in m and b:

$$\sum_{i=1}^{N} y_i = N \cdot b + m \sum_{i=1}^{N} x_i \text{ and}$$

$$\sum_{i=1}^{N} x_i \cdot y_i = b \sum_{i=1}^{N} x_i + m \sum_{i=1}^{N} x_i^2$$

Solving the set of two equations for the variables m and b yields, for summations from $i = 1$ to N:

$$m = \frac{N \sum x_i y_i - (\sum x_i)(\sum y_i)}{N \sum x_i^2 - (\sum x_i)^2}$$

$$b = \frac{-(\sum x_i)(\sum x_i y_i) + (\sum x_i^2)(\sum y_i)}{N \sum x_i^2 - (\sum x_i)^2}$$

The correlation coefficient, a measure of the degree of linearity between x_i and y_i, is given by:

$$r = \frac{\sum x_i y_i - \frac{1}{N}(\sum x_i)(\sum y_i)}{\sqrt{\sum x_i^2 - \frac{(\sum x_i)^2}{N}} \cdot \sqrt{\sum y_i^2 - \frac{(\sum y_i)^2}{N}}}$$

These equations may seem a bit of a nightmare (suitable for an appendix) but are worth working with for small data sets so that some insight is gained into the subroutines.

Exercises

The following data sets were created by taking pure analytic data (linear, quadratic and cubic) and including error using Mathcad's random number operator (rnd(x)). Apply the definitions of the best-fit slope and intercept and the correlation coefficient to each of the sets. How do you determine whether or not your data is indeed linear (Exercises 2 and 3)?

1. Linear data with random error:

x-data	1.0	2.0	3.0	4.0	5.0
y-data	-1.03	-3.92	-7.16	-9.93	-12.83

2. Quadratic data with random error:

x-data	1.0	4.0	9.0	16.0	25.0
y-data	-1.06	-10.22	-25.47	-46.25	-68.87

3. Cubic data with random error:

x-data	1.0	8.0	27.0	64.0	125.0
y-data	10.05	46.63	134.99	349.99	636.04